Lake Effect

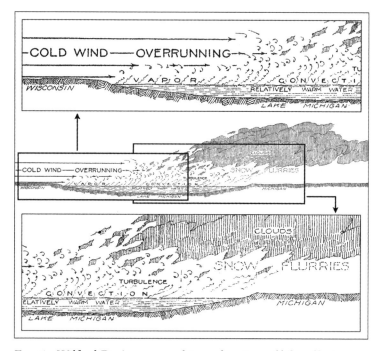

Frontis. Wilfred Day's 1921 graphic explanation of lake-effect snow.

Lake Effect

Tales of Large Lakes, Arctic Winds, and Recurrent Snows

Mark Monmonier

Syracuse University Press

First Edition 2012

12   13   14   15   16   17   6   5   4   3   2   1

∞ The paper used in this publication meets the minimum requirements
of the American National Standard for Information Sciences—Permanence
of Paper for Printed Library Materials, ANSI Z39.48-1992.

For a listing of books published and distributed by Syracuse University Press,
visit our website at SyracuseUniversityPress.syr.edu.

ISBN: 978-0-8156-1004-5

Library of Congress Cataloging-in-Publication Data

Monmonier, Mark S.
    Lake effect : tales of large lakes, arctic winds, and recurrent snows /
Mark Monmonier.
        p. cm.
    Includes index.
    ISBN 978-0-8156-1004-5 (cloth : alk. paper)   1. Snow—New York (State)—
Syracuse.   2. Winter storms—New York (State)—Syracuse.   3. Blizzards—
New York (State)—Syracuse.   4. Syracuse (N.Y.)—Environmental conditions.
5. Syracuse (N.Y.)—Social conditions.   I. Title.
    GB2601.N4M66 2012
    551.57'84274766—dc23   2012021904

Manufactured in the United States of America

This book is for my long-term co-conspirators
in the History of Cartography Project:

*Jude, Beth, Claudia, Matthew, Mary,
Joel, Karen, Peter, and Jon.*

Mark Monmonier is Distinguished Professor of Geography at Syracuse University's Maxwell School of Citizenship and Public Affairs, and editor of volume 6 (*Cartography in the Twentieth Century*) of the *History of Cartography*, funded by the National Science Foundation. He has been editor of *The American Cartographer*, president of the American Cartographic Association, and a member of advisory panels for the National Research Council and the US Environmental Protection Agency. Monmonier is author of fifteen books, including *How to Lie with Maps*; *Air Apparent: How Meteorologists Learned to Map, Predict, and Dramatize Weather*; *Spying with Maps: Surveillance Technologies and the Future of Privacy*; and *Coast Lines: How Mapmakers Frame the World and Chart Environmental Change*. For diverse contributions to cartography, he was awarded the American Geographical Society's O. M. Miller Medal in 2001, the Pennsylvania State University's Charles L. Hosler Alumni Scholar Medal in 2007, and the German Cartographic Society's Mercator Medal in 2009.

Contents

Preface

This book has simmered in the back of my mind for over ten years. Having lived thirty-five miles from Lake Ontario for nearly four decades, I've gotten to know lake-effect snow's variations and impacts, and with a long involvement in map history, including my 1999 book *Air Apparent: How Meteorologists Learned to Map, Predict, and Dramatize Weather*, I've often wondered how weather scientists used maps to figure out the unique weather that makes the downwind margins of our Great Lakes a distinctive place.

Part of this distinctiveness is the occasional snowburst, like the one I experienced years ago between Christmas and New Year's, when the university was closed and I drove over to campus to use my office computer. It wasn't snowing when I left, and I noticed nothing until Marge called to say, "It's coming down pretty fast." I looked out the window and agreed it was time to head home—a lake-effect band had apparently intensified and drifted south. Well over a half foot of snow had accumulated on what had been a bare parking lot two hours earlier. I had driven only a few hundred feet when the muffler fell off. The car now sounded like a tank, but I plowed ahead, slowly, to within a few blocks from home. Had I left ten minutes earlier I might have avoided getting stuck behind a bus that blocked the road on a small, slick hill. I backed into a grocery store parking lot—very slowly—and walked the rest of the way through the rapidly accumulating snow, which, I learned later, was falling at five inches an hour. Like most lake snow, it was light and fluffy, but shortly after I arrived home it stopped abruptly. By midafternoon, a town plow had reopened the roads, and I walked back to the market with a snow shovel, dug out the car, and drove off. I've been careful ever since to pay close attention to weather forecasts and check the online radar loop before venturing out.

Another part of this distinctiveness is gray skies and successive days with "light dustings" that contribute to impressive snowfall totals, more than three hundred inches a year in some places. Lake snow can cripple communities to the lee of the Great Lakes, but severe disruptions are rare because residents have learned to cope. For most of us, living with the "lake-effect snow machine" is a small price for our pleasant summers, delightful autumns, and comparative freedom from devastating winds and floods, not to mention the blizzards that stalk the Upper Midwest and the massive snowstorms that have paralyzed East Coast cities in recent years.

Readers with little tolerance for ambiguity might be puzzled by this picture of lake-effect snow as both an almost daily annoyance throughout the winter and a far less frequent cause of deep accumulations and hazardous whiteouts. An accurate understanding of the lake effect requires an appreciation of this paradox, with recognition that a lake-effect snowstorm, even one with thunder and lightning, rarely qualifies as a blizzard, and that so-called synoptic snow, produced by much larger weather systems that travel, is very different from lake-effect snow, which is typically lighter and is related to snowbands anchored to a large inland lake.

Lake-effect snow is also markedly different from the massive snows of the Rockies, the Sierras, and the Cascades, where moist air off the Pacific Ocean contributes to prodigious accumulations that can last well into summer. I vividly recall a fly-and-drive vacation that had to be hastily reconfigured because multiple passes through the Sierras were still blocked by snow in mid-June.

I've written this book to explain the lake effect to residents who crave a fuller understanding of causes and trends as well as nonresidents misinformed by the media stereotype of ceaselessly brutal winters. Other audiences are history buffs and weather junkies, who should relish the intriguing story of how nineteenth- and early-twentieth-century weather scientists learned to measure, map, and comprehend the phenomenon, and how their twenty-first-century counterparts are working out the likely effects of climate change on the Great Lakes snowbelts.

The story fits nicely into seven chapters, each with a tellingly concise one-word title: Recipe (chapter 1) outlines the basic physics essential to understanding what follows; Discovery (chapter 2) explores the slow cartographic recognition of lake-effect snow as a distinctive meteorological phenomenon; Prediction (chapter 3) examines the evolution of forecasting strategies; Impacts (chapter 4) limns societal effects and coping strategies; Records (chapter 5) investigates the collection and use of snowfall data and questions the national obsession with extreme weather; Change (chapter 6) looks at historical trends in snowfall and the regional implications of global warming; and Place (chapter 7) identifies seasonality as a key component in local economies and regional culture.

Lake-effect snow provides a theme for many stories, some focused on the nineteenth-century natural philosophers who didn't appreciate the phenomenon until a network of systematic observers produced data worth mapping and others highlighting their twentieth-century successors, who had difficulty making reliable short-term forecasts without satellites, Doppler radar, and computer models. Although organized networks, standardized measurements, and electronic technology figure prominently in these stories, a complete picture includes creative curiosity as well as bureaucratic inertia and false leads. Related tales reflect human adaptation to whimsical weather notorious for disrupting transportation, closing schools, canceling events of all types, and collapsing old or poorly designed buildings. Though creative coping is an essential part of the lake-effect story, I've stepped carefully around the appealing but discredited theory known as environmental determinism.

As readers acquainted with my work might expect, *Lake Effect* is also about map history, though the cartographers featured here are meteorologists and climatologists, not makers of atlases or topographic maps. On the border between natural history and social science, it integrates the history of meteorology and climatology, the history of thematic mapping, the societal impacts of weather and climate, and the regional geography (physical and human) of the Great Lakes region.

Writing for lay readers, I've avoided most of the rich but often enigmatic jargon of atmospheric scientists and operational forecasters, whose insights and hunches go well beyond what's necessary for an enlightened sense of lake-effect snow. But as weather aficionados fully understand, numbers are unavoidable, especially when discussing measurement strategies and extreme events. To make the narrative palatable to nonscientists in the United States, I've favored inches and Fahrenheit degrees but used metric units where conversion seems needless as well as awkward. I've also sidestepped the technical complexities of radar systems, upper-air soundings, and other topics more appropriate to a meteorological textbook.

Although my examples might seem biased toward Upstate New York, where I live, I have tried to provide a balanced treatment within the Great Lakes region, with particular attention to northern Michigan, the other major area for lake-effect snow. In most cases an apparent local bias merely reflects historical events. For example, the latter part of chapter 2 favors New York because the New York Board of Regents actively promoted the collection of weather data in the nineteenth century, whereas the first part of chapter 3 favors Michigan because two seminal studies from the 1920s focused on Lake Michigan. And chapter 4 is much enhanced by a revealing interview with Hank Bothwell, who worked for many years as a school superintendent in the Upper Peninsula.

A different kind of regional bias is apparent in my map excerpts, most of which have a pronounced focus on the Great Lakes. As a cartographer, I am much aware of the tradeoff between map scale and geographic scope, and make no apology for cutting off New England and the Upper Midwest, along with the rest of the country, to create room for more detail in and around the snowbelts. And because color printing would have increased the cover price, I've converted the book's few color maps to legible black-and-white illustrations.

Although I had been collecting articles and anecdotes for years, a one-semester leave from teaching responsibilities at Syracuse University's Maxwell School of Citizenship and Public Affairs jump-started the research by allowing a careful canvass of historical materials, notably the annual reports of the chief of the Army Signal Office (1870 to 1890) and the chief of the Weather Bureau (1891 to 1935), *Weather Bureau Topics and Personnel* (1915 to 1965), the *Monthly Weather Review* (1872 to the present), and the proceedings of the annual Eastern Snow Conference (1952 to the present). My research included visits to the NOAA (National Oceanic and Atmospheric Administration) Central Library, in Silver Spring, Maryland, where the legacy collection of the US Weather Bureau Library had been a key resource for *Air Apparent*. Several years ago the NOAA library scanned the Weather Bureau's serial reports, which greatly expedited my recent research. I also acknowledge the online journal collection of the American Meteorological Society, which was equally valuable for exploring twentieth- and twenty-first-century developments.

I also gratefully acknowledge the assistance of numerous informants in the meteorological community, in particular, William Angel (National Climate Data Center), Tom Atkins (chief meteorologist at WJET-TV in Erie, PA), Adam Burnett (Colgate University), Rodger Brown (National Severe Storms Laboratory, Norman, OK), Peter Chaston (author and retired NWS forecaster), Timothy Crum (NOAA Radar Operations Center, Norman, OK), Richard Grimaldi (SUNY Oneonta), Bob Muller (Louisiana State University), Tom Niziol (Buffalo office of the National Weather Service), Jessica Rennells (Northeast Regional Climate Center, at Cornell University), and the organizers of the 19th Great Lakes Operational Meteorological Workshop: Mike Evans and Mike Jurewicz (from the Binghamton office of the National Weather Service) and Mark Wysocki (Cornell University). In addition, Chris Roth (Erie, PA) shared information about the diverse career of her father, Laurence Sheridan, and James Remick (Lockport, NY) provided details about his uncle, John Remick. Mary Pedley (William Clements Library, University of Michigan) helped arrange my interview with Hank Bothwell.

I am no less indebted to numerous librarians and library staff, including John Olson, Elizabeth Wallace, and Carol Cavalluzzi (Syracuse University); Dorcas MacDonald, Betty Reid, and other staff at the SU Interlibrary Loan Office; Charlie Russo and Diane McKenney of SU's Library Delivery and Retrieval Services; and Jerilyn Marshall (Rod Library, University of Northern Iowa). Joe Stoll, staff cartographer in my department, provided sound advice about software, and the Maxwell School IT staff—especially Brian von Knoblauch, Stan Ziemba, Ed Godwin, and Mike Fiorentino—dealt promptly with system turbulence.

For insightful comments on near-penultimate versions of various chapters, I am indebted to Adam Burnett, David Call (Ball State University),

Anne Knowles (Middlebury College), Neil Laird (Hobart and William Smith Colleges), Anne Mosher (Syracuse University), Mary Pedley, and Jessica Rennells. Adam later reviewed a full final draft for Syracuse University Press.

I thank Peter Webber for introducing me to the SU Press several years ago, Alice Randel Pfeiffer and Mary Selden Evans for their strong support for my Lake Effect project, Kay Steinmetz and Fred Wellner for expert guidance through the thorny thicket of editorial production, and Lynn Hoppel for an outstanding cover. I also appreciate the vigilant eye of D. J. Whyte, my copy editor, and the support of Mona Hamlin and Lisa Kuerbis, in marketing. At home my wife, Marge, was there for me when she was needed and graciously scarce when she wasn't.

Lake Effect

1 Recipe

Syracuse, where I live, is famous (or infamous) as the snowiest large city in the United States, although residents of Buffalo, 130 miles to the west, disagree vehemently whenever Lake Erie clogs their roads with two feet of snow. What makes the two cities distinctive is lake-effect snow, a term that was new to me when I moved here in the mid-1970s. I've heard a lot about the lake effect since then, particularly after a local television station acquired Doppler weather radar in the 1990s. From November through early March in Central New York our weather is like a soap opera in which lake-effect snow is the ever-present, erratic, and sometimes even lovable protagonist. Schools close; events are canceled; cars hit poles and bring down power lines; and local lawmakers grouse about the cost of snow removal. Although days with just an inch or two of new snow are far more common, repeated "dustings" can build an impressive total snowfall.

Lake-effect snow is a distinctive type of winter weather found in relatively few places. It's native to some but not all areas bordering the Great Lakes, and it even occurs near the southeast shore of the Great Salt Lake—though smaller than Lake Ontario, the least extensive of the five Great Lakes, Utah's shallow inland sea is big enough to whip up snowfalls familiar to residents of Syracuse and Buffalo. Outside North America, lake-effect snow falls in southern Russia along the eastern shore of Lake Baikal and in northern Japan, where the island of Hokkaido borders the Sea of Japan, an appendage of the Pacific Ocean well situated to mimic the influence of a Lake Erie or Ontario.

Areas with exceptional amounts of winter precipitation are called snowbelts, a geographical term akin to Corn Belt, Rust Belt, and northern Georgia's largely forgotten Men's Pants Belt.[1] The Great Lakes are responsible for several snowbelts, but maps differ on their size and extent. While there's no official criterion for the amount of snow needed to qualify, a relatively abrupt decline at the belt's margin seems more appropriate than a rigid, largely arbitrary, round-number threshold. As I'll explain later, measuring the depth and persistence of snow is frustrated by drifting and compaction, and reliable averages are undermined by year-to-year fluctuations.

Val Eichenlaub, the geographer-climatologist who based his widely cited 1970 snowbelt map on mean annual snowfall records for 1930–60, looked for a sharp drop-off in snowfall away from the shoreline.[2] His map's smooth, almost deliberately vague dashed-line delineations (fig. 1.1) reflect the difficulty of fixing a zone that could be over fifty miles wide one year and markedly narrower the next. Like most snowbelt cartographers, he omitted outliers, which would be difficult to avoid if a rigid threshold were used.

Buffalo boosters were no doubt pleased that his Lake Erie snowbelt included their city—well, at least the southern half—while its Lake

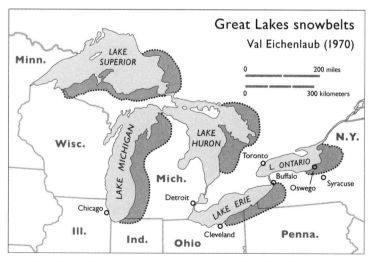

1.1. The Great Lakes snowbelts as delineated by Val Eichenlaub in 1970.

Ontario counterpart stopped about fifteen miles short of Syracuse. Buffalo's position directly on the lake no doubt helped: its average annual snowfall for the period was a mere 77 inches, far lower than the 119 inches tallied for Syracuse, apparently considered too far inland.[3] By contrast, Oswego, on the lake thirty miles northwest of Syracuse, registered 153 inches, while Fulton, not quite ten miles inland on the way to Syracuse, could boast an impressive 140 inches. I can vouch for a marked difference between Fulton and Syracuse during the time I've lived here and have no qualms about Eichenlaub's delineation. Even so, he apparently had second thoughts about Buffalo because the Lake Erie snowbelt stopped well short of the city's southern suburbs on the map in his 1979 book on the climate of the Great Lakes.[4]

A recent delineation on Wikipedia offers a more expansive view of North America's snowbelts, courtesy of a Montreal meteorologist identified only as "Pierre cb," who based his boundaries on snowfall data from the US Environmental Protection Agency and its northern counterpart, Environment Canada.[5] Instead of separate belts adjoining the southeastern shores of Lakes Erie and Ontario, a single, comparatively thick belt stretches from the Adirondacks to about thirty miles east of Cleveland, taking in Buffalo as well as Syracuse. Although the threshold was "150 cm [59 inches] accumulations or more during winter," labels were added to point out notably higher amounts within the various snowbelts. In Central New York as well as around Lake Superior, for instance, annual accumulations of 250 to 350 cm [98 to 138 inches] are typical, while the belt east of Lake Michigan racks up only 150 to 300 cm [59 to 118 inches]. Although Pierre paints a more realistic picture of areas affected by lake-effect snowfall, Eichenlaub captured the true meaning of *snowbelt* with narrower zones and separate belts for Lakes Erie and Ontario.

Snowbelt maps reveal several key ingredients in the recipe for lake-effect snow, most notably a leeward location. Because winds across the region are generally from the west or northwest, snowfall is likely in areas immediately east and southeast of the Great Lakes whenever frigid air sweeps down from Canada and picks up moisture as it crosses a large, relatively warm lake. Large inland lakes are huge reservoirs of heat that warm up slowly in summer and cool down gradually in late fall and winter.

Because the lake surface is much warmer, typically by more than 20 Fahrenheit degrees, extensive evaporation occurs when a moving air mass is in contact with the lake for a substantial distance. Although its speed across the lake might shorten or prolong the transfer of moisture, a significant lake-effect snowstorm typically requires a relatively long contact distance, or *fetch*.[6]

Fetch reflects a lake's orientation and width as well as wind direction. Lake Michigan, with a pronounced north-south trend, is sufficiently wide to support a snowbelt along the portion of its eastern (leeward) shore where westerly winds are common, while western New York's Finger Lakes, also aligned north to south, are not only too short but much too narrow for substantial lake-effect snow. (Seneca Lake, the largest Finger Lake, is only thirty-eight miles long and at most two miles wide.) Similarly, Lakes Erie and Ontario, aligned east to west, provide the fetch necessary for snowbelts along their southeastern margins.

Wind direction and the lake's alignment account for notable differences in fetch and average annual snowfall among large Great Lakes cities. For example, westerly winds reaching Buffalo (with a yearly average of 94 inches, according to recent data) must cross a much longer stretch of Lake Erie than northwesterly winds reaching Cleveland (with a mere 57 inches).[7] And while Rochester (92 inches), on the south side of Lake Ontario, occasionally receives snow from northerly winds crossing fifty miles of open water, Toronto, Canada (52 inches) enjoys a more protected position on the northwestern shore. Similarly, lake-effect snow is less frequent at Chicago (39 inches), on the southwestern edge of Lake Michigan, than at South Bend, Indiana (71 inches), twenty miles inland from the southeastern shore, where the fetch can exceed eighty miles. Because northeast winds rarely cross the lake in winter, Chicago receives almost all of its snow from more conventional weather systems associated with low-pressure cells, which typically account for more than half the snow in famous lake-effect cities like Buffalo and Syracuse.[8] If the lakes were not there, they'd still have snow in winter.

Because strong bands of lake-effect snow can penetrate more than a hundred miles inland, proximity to a large lake can be less important than wind direction. I vividly recall driving into a band of lake-effect snow from

Lake Erie when headed south from Syracuse to Binghamton. Though the area ahead looked like a long, low cloud, I quickly passed from sunny sky into a heavy snowstorm that ended abruptly about ten miles later—a whiteout is a surprise if you're not expecting it. Just to be sure, I confirmed the snow's source, more than 180 miles to the west, by looking online at the Doppler radar loop. Similarly, Toronto occasionally receives snow from northwest winds crossing Lake Huron, while Detroit (with only 41 inches of snow in an average year) is closer to the lake but too far west.

As strange as it might seem, Lake Huron can contribute lake-effect snow to Buffalo and Syracuse. A "multiple-lake interaction," as meteorologists call it, occurs when winds off one of the lower lakes (Erie and Ontario) incorporate moisture from Lake Superior or Lake Huron—perhaps both—thereby increasing the effective fetch.[9]

If winds and fetch were the only factors, Buffalo, near the easternmost tip of Lake Erie, would receive at least as much snow as Oswego, not quite at the east end of the appreciable shorter Lake Ontario. But Lake Erie's greater size cannot offset an average depth of only 62 feet, which makes it the shallowest of the Great Lakes. With the lakebed a mere 210 feet below the surface at its deepest point, Lake Erie can freeze rapidly, reducing the moisture available for lake-effect snow.[10] Ice begins to form in December, and 94 percent of the lake freezes over in most years, usually by mid-February.[11] Because the thickness of the ice is important, extensive ice cover will not completely stifle lake-effect snow—if the fetch is right and the ice relatively thin, noteworthy snow can accumulate.[12] By contrast, Lake Ontario, with average and maximum depths of 282 and 802 feet, respectively, stores much more heat and remains largely open to evaporation throughout the winter. In most years, no more than 21 percent of its surface freezes over.

Not all years are typical, though, and Buffalo is particularly vulnerable when Lake Erie is slow or late in freezing. Not surprisingly, the heaviest snows occur when the lake is relatively warm and evaporation more intense. Normally November and December register only 9 and 20 inches of snow, respectively, but on the six occasions since 1940 that Buffalo's total snowfall exceeded 130 inches, the snowiest month was usually November or December.[13] During the 2001–2 season, for instance,

December alone (with 83 inches) accounted for nearly two-thirds of the seasonal total (132 inches). And when markedly more snow (159 inches) fell a year earlier, November (46 inches) and December (50 inches) jointly contributed 60 percent. In 1995–96 and 1983–84 December snowfalls (61 and 52 inches, respectively) were equally prominent factors in substantial seasonal totals (141 and 133 inches). During the 1977–78 season, December (53 inches) and January (57 inches) together constituted over 70 percent of the total (154 inches), and a similar collaboration (61 and 68 inches) explained the previous season's massive total (199 inches)—more than twice the normal amount.

Exceptional monthly snowfalls in Buffalo usually reflect a single massive "lake-effect event." Forecasters at the local National Weather Service office give their storms unique names that mimic the alphabetical progression of personal names like Ana, Bill, and Claudette, used to identify hurricanes. Instead of alternating male and female names and mixing in a few Dutch, French, and Spanish monikers to appease Europeans, the NWS Buffalo staff adopts a different naming theme every season. In 2000–2001, when the theme was trees, the most memorable storm was the season's third, named Chestnut, which raged from Monday, November 20, through Thursday the 23rd, Thanksgiving.

Dubbed the "Millennium Snowburst," Chestnut was a category 5 storm on the "Lake Flake Scale," a subjective intensity index that Buffalo forecasters modeled after the Saffir-Simpson scale for hurricanes and the Fujita scale for tornadoes. In contrast to a "wimpy" one-flake storm that's easily shoveled aside, a five-flake "epic mega-storm" lasts several days, stalls transport, closes schools, and stifles business in general. Chestnut was the only five-flake storm that season, and one of only six logged between late 1998 and early 2009 for a forecast area covering eastern Lake Erie and all of Lake Ontario.

Described as a "wintertime flash flood" with "abundant lightning" lasting eight hours, the storm not only immobilized Buffalo with over two feet of snow from Lake Erie but dumped 28 inches of Lake Ontario snow on the village of Parish, twenty miles north of Syracuse.[14] Flakes began falling in Buffalo shortly after noon on Monday, after a cold front accompanied by southwest winds set up a strong band of lake-effect snow over

Lake Erie. At 5 p.m. the Buffalo radar showed a snowband roughly twenty-five miles wide stretching over 180 miles from the middle of Lake Erie across western New York into Lake Ontario north of Rochester (fig. 1.2). Snowfall was most intense just east of Buffalo and a bit farther east, on the Erie County line, as shown by the darkest areas on my black-and-white map (bright yellow on the original radar image). The timing was disastrous for workers and schoolchildren sent home too late to avoid being trapped in stores, cars, and stalled school buses by snow falling at 2 to 4 inches per hour. The band remained in this position until 9 p.m. before disintegrating. Lake-effect snow resumed on Tuesday and Wednesday, when westerly and northwesterly winds brought Lake Erie snow to areas south and southeast of Buffalo and Lake Ontario snow to Oswego and Syracuse. As the wind shifted, the bands changed direction and moved north to target new areas into the wee hours of Thanksgiving morning. For many residents the storm was memorable for having disrupted travel plans or holiday food shopping.

As loyal viewers of the Weather Channel or local TV weathercasts are well aware, radar maps delineate lake-effect snowbands with vivid colors.

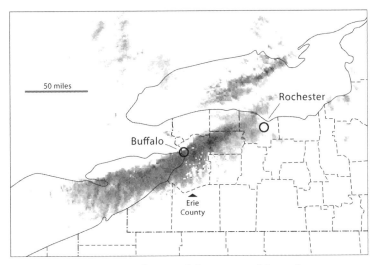

1.2. Black-and-white rendering of the 5 p.m., November 20, 2000, Buffalo radar image shows the "Chestnut" storm centered over Buffalo and Erie County.

A rotating radar beam seeks out particles of condensed water (including ice crystals) that are large and dense enough to reflect the signal back to the antenna: the pulse's round-trip travel time provides an estimate of distance while the particle's reflectivity helps distinguish hailstones from snow flurries or light rain.[15] Although the meteorologist can vary the association of reflectivity and color, the map I converted to black-and-white for figure 1.2 used a blue-green-yellow-orange sequence, whereby a light blue at a snowband's edge represents tiny, barely detectible ice crystals, hundreds or thousands of feet in the air, which can evaporate before hitting the ground. Although snow might not be falling in the blue areas, the bright yellow blotches toward the center of the band typically signify larger particles able to reflect much more of the beam's energy from areas where snow most certainly is falling, perhaps at rates of several inches per hour. In between are zones with progressively less moisture—frozen, of course— represented in turn by dark green, medium green, light green, dark blue, and medium blue. Although orange splotches can appear within a yellow region if the snowfall is exceptionally intense, hail and sleet, which are highly reflective, rarely accompany lake-effect snow.

As lightning accompanying the Chestnut storm suggested, lake storms are somewhat similar to summer thunderstorms. Both depend on saturated air, which rises rapidly when condensation produces raindrops or snowflakes. Two principles from middle-school science class explain what's happening. The first relates density to temperature: warm air is lighter than cold air and thus rises, whereas cool air is heavier and falls. The second involves the energy required to evaporate water. In the same way that vapor rising from a pan of water on a stove absorbs heat from the burner below, the water evaporating from a relatively warm lake needs "latent heat" to keep its molecules apart. When moisture in the air converts to rain or snow, heat energy is released and the air becomes warmer, and because it's now relatively warm, it rises. What's more, because the uplifted air is now a bit cooler than its immediate surroundings—air temperature generally declines with increasing altitude—still more condensation occurs, which in turn triggers more warming and lifting. Get the picture? Highly humid air is considered "unstable" because a nudge upward can trigger an updraft likely to produce precipitation.

Warming of the air in direct contact with the lake surface also enhances lifting. Meteorologists call this "heat flux" or conduction. In winter a frozen lake can still warm frigid air passing over the ice. In late summer or early fall, when temperature is too warm for snow, precipitation can fall as lake-effect rain.

In much the same way that summer thundershowers don't occur everywhere at once, lake-effect snow is not a single vast conveyor belt moving moisture steadily eastward as cold air crosses a warm lake. Thunderstorms and snowbands share a vertical structure that concentrates precipitation within a limited area where rising currents of warming, buoyant, condensing air produce falling particles of rain, snow, sleet, or hail. T-storms typically occur in late afternoon or evening, after bright sunlight has warmed the land surface, which in turn heats the air immediately above, triggering a rapid updraft of relatively warm, humid air. Condensation is apparent in the rapid formation of a dark, moisture-laden cloud extending upwards perhaps seven miles or more. Although thunderstorms vary in size, complexity, and duration, the area receiving rain at any one time from a simple, single-cell thundershower, with just one main updraft, is a more or less circular area only a few miles across.

Lake bands are similarly convectional, with the updraft most pronounced along the axis of the band, which is comparatively longer and flatter than a thunderstorm cell. While some snowbands are small, perhaps only two miles wide, bands over twenty miles wide and more than a hundred miles long can produce memorable snowfalls. A snowband typically becomes broader on its downwind end, as a result of longer exposure to the warm lake surface, and its convective structure grows vertically as well, extending upward perhaps a mile and a half or more.[16] And unlike the single-cell thundershower that forms and disintegrates within a half hour, a well-organized lake band that lasts a half day or longer can remain in place or change position and direction.[17]

Some lake-effect storms have multiple bands, which usually occur when wind direction deviates by more than 30° from the direction of maximum fetch. Air flowing roughly parallel to the long axis of the lake can take full advantage of a long fetch and produce a single long, comparatively narrow band with intense snowfall and even lightning, as occurred during the Chestnut storm. By contrast, multiple bands form when airflow is perpendicular to the lake's axis and a relatively short fetch produces less intense lake snow. Westerly winds on Lakes Ontario and Superior and westerly to southwesterly winds across Lake Erie produce the most intense lake-effect snows, while on Lakes Huron and Michigan a westerly wind will form multiple bands, which don't reach as far inland and bring less snow to a wider area. If the prevailing flow changes during a storm, the type of band will change as the fetch grows or shrinks.

Weather satellites afford an overhead view that illustrates the broader influence of pressure systems on snowbands. Figure 1.3, captured mid-morning on the second day of the Chestnut storm, shows the effect of northwesterly winds from Lake Huron and Georgian Bay as well as westerly winds from Lakes Erie and Ontario. Light areas representing sunlight reflected from cloud tops outline several snowbands, including one delivering Lake Erie snow to areas just south of Buffalo and two distinct bands bringing Lake Ontario snow to areas east of the lake. The curved pattern of the snowbands represents air moving counterclockwise around a low-pressure system centered up in Quebec, outside the area shown. Pressure systems able to whip up winds with a long, productive fetch are an important ingredient in the lake-effect recipe

Another ingredient is surface friction. Although snow falls offshore, the amount and intensity increase when a snowband extends from the flat, relatively smooth lake surface onto an uneven landscape where hills, gullies, and woodlands impede movement. This causes the air to pile up, creating turbulence and upward movement. Think of a crowded department store in which shoppers don't step forward quickly when an escalator delivers them to a new floor. Lake-effect snow can be just as chaotic when unstable air, rich in moisture and freshly heated by a relatively warm lake, encounters comparatively rough and cold terrain, which triggers condensation, convection, and heavy snow. As the snowbelt map (fig. 1.1) confirms, the effects of this spatial change in surface friction near the shoreline can extend inland twenty miles or more.

Curious about the impact of the Chestnut storm, Buffalo forecasters mapped the amount of new snow from the three-day event (fig. 1.4). Using reports from volunteer observers as well as radar images updated every five

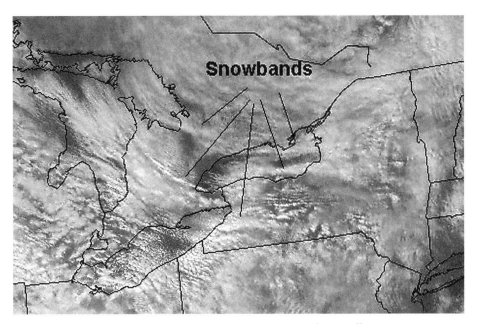

1.3. Cloud cover image captured by the GOES-8 weather satellite at 10:15 p.m., November 20, 2000, shows snowbands from Lakes Huron, Erie, and Ontario delivering lake-effect snow to the areas south of Buffalo and east of Lake Ontario.

minutes, they delineated areas in New York State receiving at least three inches of snow within range of their radar tower in the Buffalo suburb of Cheektowaga and the next National Weather Service radar site to the east, in Montague, New York, a remote town twenty-five miles due east of Lake Ontario. The map's four snowfall categories, the largest representing more than two feet of snow, reflect the narrow, intense band that targeted Buffalo early in the storm, on Monday afternoon and evening, as well as additional bands that had moved into place the following morning, as shown on the satellite image (fig. 1.3). Because the zones with the most snow are not immediately adjacent to the lake, the map reveals a significant but optional lake-effect ingredient: the lifting influence of highlands like Tug Hill, east of Lake Ontario, and the Appalachian Highlands in western New York, south of Buffalo. Heavy snow occurs when moist west winds off the lake encounter rugged terrain, which forces unstable air to rise abruptly.

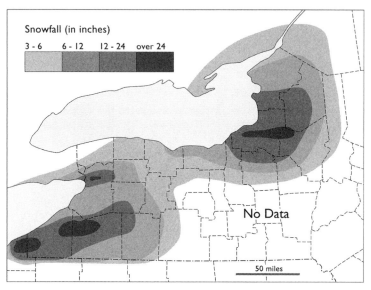

1.4. Total snowfall from the lake-effect storm Chestnut, November 20–23, 2000, as estimated by the Buffalo, New York, office of the National Weather Service.

In the meteorologist's lexicon, the enhanced snowfall squeezed out by highlands in the snowband's path is *orographic* precipitation, a term derived from *oros*, the Greek word for mountain. In the United States orographic rain and snow are especially prominent in the Pacific Northwest, where mountain ranges in the path of moisture-laden air off the Pacific Ocean produce over 100 inches of precipitation annually on the west-facing slopes of the Coast Range and the Cascades. With well over twelve feet of rain, Washington State's Olympic Peninsula has several pockets of rain forest, a phenomenon normally associated with the tropics. Farther east, the orographic effect accounts for the snow-covered peaks in the Sierra Nevada ("snowy mountain range" in Spanish) and the Rocky Mountains, home to the nation's highest snowfalls. By contrast, precipitation maps show a "rain shadow" on the east-facing slopes, which are comparatively dry because the topographic hurdle has rung most of the moisture out of the Pacific air. Detailed snowfall maps reveal a similar "snow shadow" east of Tug Hill as well as in numerous valleys throughout western and central New York.[18]

Tug Hill endures some of the heaviest snows east of the Rockies. Extending forty-five miles inland from Lake Ontario's eastern shoreline and bounded at the north and south by the latitudes of Watertown and Oswego, it catches nearly the full range of westerly snowbands. Its 2,100 square miles are sometimes mistaken for an outlier of the much larger Adirondack Mountains, farther east, but the latter's 5,000 square miles of more rugged terrain and higher peaks are underlaid by igneous and metamorphic rocks, whereas the Tug Hill was carved from sedimentary rocks, mostly sandstone. It is sometimes called the Tug Hill Plateau, which is misleading because it's not a comparatively level tableland like the Columbia Plateau. Instead, Tug Hill is a tilted upland with elevations rising from around 250 feet near the shoreline to 2,100 feet near its eastern edge—like a table tilted upward at one end, Tug Hill forces air from the west to rise steadily, spreading lake-enhanced orographic snowfall over the bulbous protrusion prominent on the maps of Great Lakes snowbelts (fig. 1.1) and total snowfall from the Chestnut storm (fig. 1.4).[19]

Orographic lifting is an important factor in other Great Lakes snowbelts. Careful examination of the shaded relief map in figure 1.5 reveals a significant topographic influence in all or part of the six heavy-snow zones. What's more, most snowbelts have key physiographic constituents with distinctive names; examples include the Keweenaw Peninsula, on Michigan's Upper Peninsula south of Lake Superior; the Michigan High Plains, northeast of Lake Michigan on the state's Lower Peninsula; the Appalachian Highlands, near the southeastern margin of Lake Erie; and Tug Hill, east of Lake Ontario. The massive Adirondacks, farther east, are usually too far from the shoreline for massive amounts of lake-effect snow. But lots of snow falls here, courtesy of East Coast storms pumping in moisture from the Atlantic.

Research papers, textbooks, and websites offer diverse recipes for lake-effect snow. Like handwritten heirlooms and cookbook formulas for everyday delights like apple pie, their ingredients are largely similar—try making apple pie without apples and some form of flour. Almost everyone's list includes a large, relatively warm lake, cold winter winds, and a lengthy fetch, but requirements vary as widely as scratch bakers differ from cooks satisfied with store-bought pie shells.

1.5. Shaded relief map produced by the Jet Propulsion Laboratory using data from NASA's Shuttle Radar Topography Mission.

Some writers specify a minimum fetch, others do not. A *USA Today* article on lake-effect snow's "many ingredients," demands a fetch of "at least 60 miles," while the more tolerant *Encyclopedia of Earth* requires "at least 80 kilometers," a metric distance a shade short of fifty miles.[20] Snowbelt guru Val Eichenlaub might be read as endorsing the sixty-mile minimum in his book on Great Lakes climatology, but his diverse examples mostly demonstrate that longer is better.[21]

Even so, a long fetch will not guarantee a big snow. Laurence Sheridan, the New York college professor who set out perhaps the earliest lake-effect recipe in a 1941 article on Lake Erie, included "westerly winds" in an ingredients list lacking both fetch and a minimum distance.[22] Nine years later, Buffalo forecaster Barney Wiggin included "a long fetch over water" as one of five key ingredients in a meteorologically concise list that emphasized "a strong flow of polar air" and "a large temperature differential between the air and water."[23] Significant lake snowfall, he further observed, is favored by a long, comparatively narrow "longitudinal" cell, certain to have the requisite fetch.

Fetch is more a topping than a main ingredient, according to Buffalo's chief meteorologist, Tom Niziol, who weighed in with a "decision tree" based on "yes" or "no" answers to four questions framed specifically for Lakes Erie and Ontario.[24] His first question—"Is the temperature difference between the lake and the 850 mb level 13°C [23°F] or more?"—underscores the importance of a relatively warm lake, obviously not covered with thick ice. His "850 mb" (850 millibars) pressure level represents an altitude of roughly 5,000 feet, where the air above exerts only about 5/6 the pressure at the surface. Niziol then asks whether winds near the surface and at altitudes of about 5,000 feet are blowing from the southwest, west, or northwest across Lake Erie or from the northeast, north, northwest, west, or southwest across Lake Ontario.[25] Experience shows that winds from other directions will not produce lake-effect snow near Buffalo. His last two questions ask about "wind shear," in particular the difference in wind direction near the surface and at about 10,000 feet. If the wind directions are between 30° and 60° apart, lake-effect snow is merely "possible," but if the difference is less than 30°, lake snow is "likely." Unless wind shear interferes, cold air streaming in a favorable direction across a relatively warm lake will provide sufficient fetch for a snowfall rate of an inch or more an hour.

Jeff Haby, the former college meteorology professor who runs the website *theweatherprediction.com,* also likes questions with "yes" or "no" answers. His recipe for "heavy lake effect snow" reminds me of a self-scoring quiz for predicting longevity or success as a marriage partner: there's no critical score, but the more yeses, the better.[26] Seven moderately technical questions address not only wind shear and temperature difference but also the "depth of arctic air," whether the air within a mile or so of the surface (within the "planetary boundary layer") is below freezing, and whether the wind speed is between 10 and 45 miles per hour. Gentle winds concentrate the snow near the shoreline, whereas overly strong winds allow too little evaporation. Haby mentions fetch directly—"greater than 100 kilometers [62 miles]" is good—and also notes the boost from "significant orographic lifting...downwind from the lake."

Applicable throughout and beyond the Great Lakes, his recipe's diverse list of ingredients suggests that lake snow is more like bouillabaisse than chicken soup.

How far beyond the Great Lakes region does lake-effect snow occur? I've already mentioned the Great Salt Lake, which lives up to its name in size and salinity with a generous expanse of open water all winter long despite a maximum depth, even in relatively wet years, of less than 50 feet. Other inland water bodies known to produce lake-effect snow include Hudson Bay, Great Bear Lake, Great Slave Lake, Lake Athabasca, Lake Winnipeg, and Lake Nipigon, all in Canada; Lake Baikal, in Russia; and even Lake Tahoe, on the Nevada-California border. Snowbelt cartographer Val Eichenlaub, who believed that heavy lake-effect snowstorms were largely a Great Lakes phenomenon, recognized the Sea of Japan as a source of snow for the Japanese islands of Honshu and Hokkaido as well as a similar situation east of the North Sea in the Baltic States. Lake snow was "notably absent" in the Southern Hemisphere, he vaguely explained, because "geographic and atmospheric factors do not combine."[27]

I searched for reports of lake snow downwind from the Caspian Sea— the world's largest inland lake on some lists, a small ocean on others (because of its salt)—but found nothing. But serendipity intervened, and my search turned up an article in the journal *Natural Hazards* titled "A Severe Sea-Effect Snow Episode over the City of Istanbul"—during seven days in February 2005 northeasterly winds off the Black Sea dumped nearly two feet of snow on the Turkish capital.[28] The author's quest for a cause uncovered key ingredients of Great Lakes snowstorms.

Recent studies suggest that ocean-effect, bay-effect, or sea-effect snow—whatever it's called—is fairly widespread, thanks to cold winds, salt water's resistance to freezing, and in some places, relatively warm ocean currents. Not as frequent or consistent as its Great Lakes counterpart, an ocean-effect snowfall rarely exceeds a couple of inches. Even so, local forecasters must be wary of cold polar air masses moving onshore—in January 1999 an ocean-effect storm dumped as much as a foot of snow on the Boston suburbs.[29] In addition to the North Sea and the Sea of Japan, mentioned earlier, ocean-effect snowstorms have been reported near the mouths of Chesapeake and Delaware Bays, New York Harbor,

Massachusetts Bay, the Bay of Fundy, and the Gulf of St. Lawrence. I've
added these locations to my map of "lake-effect" snow sightings (fig. 1.6).

What my map does not reflect are recent reports of lake-effect snowbands over Lake Champlain, on the New York–Vermont border, and the
Finger Lakes, in Upstate New York. Sharing a largely north-south elon-
gation, they're all narrow and much smaller than the Great Lakes, and
too meteorologically insignificant to support anything worthy of the name
snowbelt. Even so, Neil Laird, who teaches at Hobart and William Smith

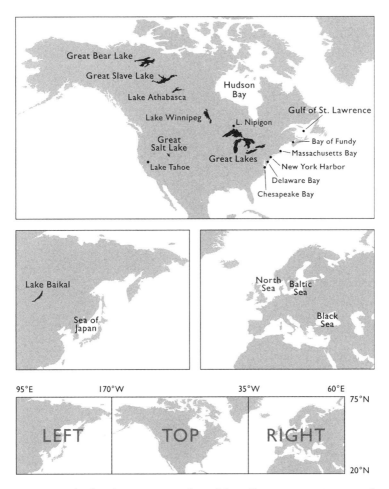

1.6. Water bodies known to produce lake-effect snow or its coastal
counterpart.

Colleges, in Geneva, New York, at the top of Seneca Lake, has documented numerous instances of lake-effect and lake-enhanced snow in the region. A radar-based study reporting 125 "lake-effect events" in the Finger Lakes between 1995 and 2006 confirmed some clearly defined snowbands with downwind deposition as well as mere local "lake enhancements" of larger regional snowstorms.[30] What's more, the *Monthly Weather Review* featured radar images of "A Lake-Effect Snowband over Lake Champlain" as the "Picture of the Month" in its November 2007 issue—a more lasting distinction perhaps than the lake's two and a half weeks as the sixth Great Lake in a 1998 ploy to qualify Vermont for Sea Grant funding.[31] However fluffy and shallow, occasional lake snow in Ithaca, New York, downwind from Cayuga Lake, demonstrates the recipe's power when polar winds align with an unfrozen, relatively warm lake. Discovery of these small, fleeting snowbands demonstrates the value of sensitive instruments and scientific curiosity.

2 Discovery

Before I convinced myself that the story of lake-effect snow deserved a full-fledged book, I pored over some old texts to see what nineteenth-century geographers knew about snowbelts. What I found was both disappointing and puzzling. For example, in their 1847 *Geography of the State of New York*, Joseph Mather and Linus Brockett touted the importance of Lakes Erie and Ontario as waterways and noted the threat to lake navigation of "storms of great violence," but they ignored lake snows entirely, even when discussing Buffalo and Oswego County.[1] Equally clueless was John Homer French, who reported the lakes' size and depth in his 1860 *Gazetteer of New York State* but sidestepped their impact on climate.[2]

Particularly surprising was the skimpy treatment of snow of any type in the 113-page *Climate of the State of New York*, written by New York's official state meteorologist, Ebenezer Tousey Turner (1862–1942), and published by the State Assembly in 1894.[3] Turner devoted less than two pages to snowfall, for which relevant data were "very meager for the State as a whole." Aware of "very heavy local amounts . . . in the southwestern counties, especially in the vicinity of Lake Erie and . . . portions of Lewis, Oneida, and Madison Counties, where the total snowfall is generally the greatest to be found east of the Rocky Mountains," he made no connection with warm lakes and polar air. Turner was even more maddening in his 1900 *Journal of the American Geographical Society* article "The Climate of New York," which also served as the chapter on climate in Ralph Tarr's 1902 *Physical Geography of*

New York State, aimed at geography teachers.[4] In a section headed "Influence of Great Lakes," he noted that "the northwesterly winds of winter, in passing over the lake, are raised to a temperature considerably higher than obtains on the north shore" but said nothing here about snow, despite observing that "this influence is felt throughout the portions of the State lying to the south and east of the lake."

That Turner saw no need to elaborate on "this influence" might reflect both the local impact of lake-effect snow and its generally mild persistence—just a little snow, albeit for weeks in a row at times—which was quite unlike the huge cyclonic snowstorms that invaded the Middle Atlantic states and New England once or twice in a typical year. A national weather service had been operating since 1870, but its forecasters were focused on broad, *synoptic-scale* features like hurricanes, northeasters, and other large cyclonic storms, several hundred miles across and described by a center of low pressure that could be tracked on a map and forecast a day or two ahead. In winter, these storms could bring heavy snow, strong winds, and disruptive drifting—all the hallmarks of a blizzard—but Buffalo and Oswego were usually too far north to share the misery with Baltimore or Philadelphia. What passed for weather science in the mid–nineteenth century was still struggling to monitor smaller synoptic-scale storms, like the fast-moving Alberta clippers, which bring snow to the Great Lakes region and the Northeast, and can intensify the lake effect. Although lake snow can mimic a blizzard on occasion, as during the Chestnut snowstorm, meteorologists had no way to map snowbands, track a lake storm's development, or survey its impact systematically. Lake snowstorms are *meso/micro-scale features*, best understood with high-resolution weather radar and the God's-eye view of satellite imagery.

Networks for collecting weather data evolved slowly from isolated seventeenth- and eighteenth-century observers committed, almost obsessively, to looking up and jotting down. Concerned less with curiosity about the atmosphere than with weather's effects on agriculture or health, their measurements and weather diaries provided a map of dots that no one bothered to connect. George Washington and Thomas Jefferson, who owned plantations, kept weather journals that demonstrated little more than an appreciation of numbers and systematic observation.[5]

Charleston, South Carolina, physician John Lining (1707–1760) hoped his daily measurements of air temperature, pressure, and rainfall, recorded conscientiously from 1737 through 1753, might reveal a link between weather and epidemics of yellow fever.[6] Seeing no definitive correlation and unaware of the role of insect vectors, he concluded that the disease was somehow infectious. Not utterly pointless, published weather journals reflected an understanding of measurement error and the need for consistency and precision.

Good questions and promising hunches emerged only after natural philosophers (as early nineteenth-century scientists were called) began to collect data, posit explanations, and publish their theories in the *Transactions of the American Philosophical Society* and similar periodicals read by like-minded investigators. William Redfield (1789–1857), who made saddles and promoted steam navigation for a living, was a notable pioneer. While traveling through southern New England in 1821, he discerned the circular motion of winds in severe storms after noting that fallen trees in various parts of the region were pointing in different directions. After adding data points from observers as far away as the West Indies, he published his findings in the *American Journal of Science* in 1831.[7] Redfield refined his "whirlwind" theory, which blamed severe storms on gravity and the earth's rotation, in a 1843 *Transactions* article, which fueled a lively debate with James Pollard Espy (1785–1860), who contended that storms were convective systems with winds rushing inward toward their center, where air was rising because of heat from below.[8] As often happens, both explanations were partly right.

A popular author and lecturer who had taught school and practiced law earlier in his career, Espy had his own network of observers. He lived in Philadelphia, where he chaired the Joint Committee on Meteorology, established at his urging in 1834 by the Franklin Institute and the American Philosophical Society. The Institute's *Journal* published his monthly summary of weather observations, contributed by a coterie of volunteers, whose ranks had grown to 110 by 1842—but only 50 had barometers.[9] Drawing on reports for June 20, 1836, when a huge storm brought rain to much of the country's midsection, Espy found support for his "centripetal" theory with a map summarizing wind direction at eighteen locations between eastern Massachusetts and western Ohio. Acclaimed by science historians as the first synoptic-scale cartographic snapshot of American weather, the map debuted in the *Journal* in 1837 as a crude wood cut (fig. 2.1) and appeared again, four years later, in Espy's *Philosophy of Storms* as a spiffier copper-plate engraving.[10]

Aware that meaningful data demanded a dispersed corps of observers with uniform, reliable instruments, Espy appealed to the Pennsylvania Legislature and the US Congress for support and proposed a scheme for overcoming periodic droughts. Although few senators considered his theory a useful recipe for rainmaking—start a big burn, he proposed, and let convective rising and inward-bound moist air do the rest—Espy's lobbying led to dual appointments in Washington in 1942, as a professor of mathematics at the Navy Department and as meteorological advisor to the Surgeon General's Office at the War Department. The army clearly understood Espy's vision for a national network of weather observers. In 1814, during the War of 1812, Surgeon General James Tilton (1745–1822)

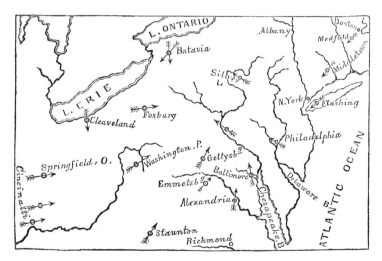

2.1. Espy's winds map for the June 20, 1836, storm. The Joint Committee's 1837 report in the *Journal of the Franklin Institute* encouraged close examination of the tiny arrows: "By casting the eye on the wood cut, it will be seen at a glance that the wind blew on all sides towards the point of greatest rain."

had ordered post surgeons to keep weather diaries and forward reports to Washington. Although Tilton was concerned primarily with the effects of weather and climate on health, data collected by the Army Medical Department provided a foundation for the systematic study of American climatology, especially in the West.[11] His successors expanded the endeavor and published substantial multiyear compilations in 1826, 1840, 1851, and 1855. In the 1817 the General Land Office also started to collect weather data, but the effort collapsed when its principal supporter died in 1822.

Although the military culture of command and control gave the Surgeon General a cadre of disciplined observers, army posts were too widely dispersed to detect meaningful local patterns. To address this need, at least for part of the country, Simeon DeWitt (1756–1834) asked private academies throughout New York State to collect weather observations using thermometers and rain gauges provided by the state. Well known to map historians for his work as a geographer and surveyor, most notably for his detailed 1802 map of the state, DeWitt was vice-chancellor of the New York Board of Regents, which oversaw the state's schools and colleges. When the regents ask, principals comply, at least those eager for a share of the state's "Literature Fund." DeWitt chaired a three-person committee that provided instructions for making, recording, and reporting observations, including number of days with snow and prevailing wind direction (separately for morning and afternoon) to the nearest of the eight half-quarter compass points. The system survived until 1863, when the state legislature, distracted by the Civil War, withdrew funding.

In 1855 the regents released a 518-page summary of the committee's annual reports for 1826 through 1850.[12] Although "a considerable portion but not all" academies had cooperated, the results were adequate for a map of tiny arrows illustrating prevailing wind direction at the sixty-two sites—the report's only map. My much smaller rendering (fig. 2.2) shows winds primarily from the west and a concentration of cooperating academies near transportation corridors like the Hudson River and the Erie Canal.[13] Few were even near a snowbelt.

The regents' snow-day counts, which might have picked up the lake influence, were too inconsistent for their report's "General Summary."

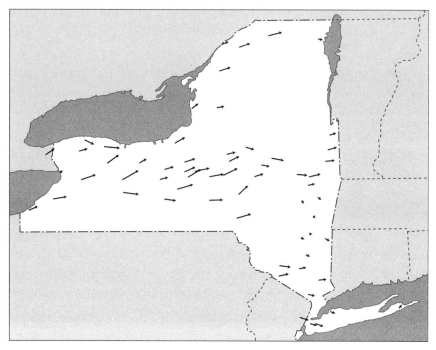

2.2. Condensed version of map in the 1855 *Meteorological Observations* showing prevailing wind direction at academies reporting to the New York Board
of Regents. Each arrow's point represents station location, its direction represents dominant wind direction, and its length represents that direction's relative
prevalence.

Observers had been asked to record type of precipitation (rain, snow) for
whole days and sky conditions (clear, cloudy) for half days, but most of
them had apparently misunderstood the instructions and recorded rain
and snow by half days. In perhaps the first published indication that measuring snowfall is not straightforward, "this want of uniformity" explained
why "the mean or total number of days on which it rained or snowed were
not ascertained or entered in the foregoing tables."[14]

Although the Army Medical Department had tallied snowfall days
since 1818, and the Franklin Institute since around 1836, these observations added little to periodic reports focused on temperature, wind, and
precipitation, typically listed as "rain and melted snow."[15] In the early

years army surgeons didn't even measure precipitation—then as now, rain gauges were more expensive and cumbersome than thermometers and wind vanes, but counting days with rain or snow required only a notebook and one good eye. The Surgeon General's 1826 report consists largely of forty-eight month-by-month summaries for the years 1822 through 1825. In addition to listing a variety of temperature and wind observations, these monthly summaries report number of days for four kinds of weather: fair, cloudy, rain, and snow. According to a consolidated table for the whole period, four of the eighteen forts reported no snow, and another four averaged less than one snow day a year.[16] Fort Brady, in Sault Ste. Marie, at the eastern end of Michigan's Upper Peninsula, was the snowiest, with a yearly mean of seventy-two snow days, which suggests that the army surgeons dutifully recorded mere dustings as well as blizzards.

By 1855, when the Surgeon General's Office issued a report covering the previous twelve years, measurements and tabulations were more comprehensive and exact, thanks to instructions drafted by the Army Medical Board after consulting Espy and the New York Regents, among others. "At every fall of rain, snow, hail, or sleet, the time of its commencement and end will be recorded, and the quantity which fell, as indicated by the rain-gauge."[17] The report described a conical rain gauge, graduated in hundredths of inches for the first three tenths, and in tenths and half-tenths thereafter—precise, expensive, and worth protecting from expanding ice. "In freezing weather, when the rain gauge cannot be used out of doors, it will be taken into the room, and a tin vessel will be substituted." Accuracy required careful placement, "where no drift snow be blown into it." Depth of snow in the gauge was deemed irrelevant because "during a continued snow storm, the snow may be occasionally pressed down." To ensure consistency, "the contents of the vessel must be melted by placing it near a fire, with a cover to prevent evaporation, and the water produced poured into the gauge to ascertain its quantity." Besides reporting the numbers of days of rain and snow, separately, the tables now listed total precipitation, in inches. Fort Brady was still the snowiest outpost, with a tally of snow days ranging from 33 in 1847 to 71 in 1853, but after the snow was melted its average annual precipitation was an unremarkable 31.35 inches.[18]

Not every observer bothered to melt the snow. As the paragraph introducing rainfall tables for the individual forts disclosed, "In some cases the amount falling in snow is not fully given as water, and some such omissions have been supplied by taking one-tenth of the reported depth of snow as its equivalent in water."[19] While conceding "some general deficiency in the measurements of the water falling in snow," the report gave the 10:1 ratio a grudging endorsement, calling it "sufficiently near to accuracy for any general purpose, though for the southern latitudes it would give too small, and for extreme northern districts too great a quantity of water."[20] Having lived in Maryland and Central New York, I can confirm that southern snow is generally wetter, heavier, and more difficult to shovel than northern snow, especially lake-effect snow, which is often (but not always) light and fluffy. Although observers have reported snow-to-liquid ratios ranging from less than 2:1 to more than 40:1, a recent study of fresh, noncompacted snow throughout the United States suggested 13:1 as a generally reliable national average. More revealing, though, are regional means ranging from 7:1 in South Carolina to 16:1 in northern Michigan and parts of the Rockies.[21] Within the Great Lakes snowbelts, ratios between 14:1 and 16:1 are typical.

Unlike the 1826 *Meteorological Register*, which interspersed tables of observations and measurements with tedious, unperceptive discussions of the numbers, the 1855 Surgeon General's report looked beyond data based on an idiosyncratic distribution of army bases to point out regional and local differences in precipitation and to posit explanations. A few of these interpretations even mention Great Lakes snow. In discussing annual rainfall, the report's author reasoned, "If the quantity of water precipitated at all seasons depended on local sources of supply in the evaporation from ocean or lake surfaces, there should be . . . a considerable increase on the hills in the vicinity of the lakes in winter or the colder seasons. There is, undoubtedly, a partial or small increase of this sort, which is most conspicuous in New York and in the vicinity of Lake Superior." Without mentioning lake-effect snow by name—the term was rarely used before the 1950s—he identified a key ingredient, namely, "the greater temperature of the lakes at this season, and . . . their slower cooling [whereby the] local atmosphere retains a degree of humidity and a capacity for moisture which does not belong to the atmosphere of the land areas in their vicinity."[22] And in discussing autumn precipitation out west, in the Great Basin, he observed, "The snows are early, also, occurring quite as soon as in the lake district of the east."[23] A few pages later, he identified an orographic influence in "the elevated districts near the lakes and toward the Atlantic, as in the highlands of New York and New England, [which] are most abundant in snow."[24]

These insights reflect the thinking, and most likely the pen, of Lorin Blodget (1823–1901), considered the father of American climatology. In the report's preface, Assistant Surgeon Richard Coolidge acknowledged the help of Blodget, who was "associated with me in the preparation of the entire work."[25] In sharing credit Coolidge also encouraged exploitation: "The isothermal and rain charts were designed and prepared exclusively by [Blodget], and he is entitled to whatever of scientific value may attach to them . . . and the results which they exhibit." Blodget's charts, "which have been prepared exclusively from data in the Surgeon General's office," were "the feature which distinguishes this register from its predecessors." As far as I can tell, the rainfall charts—one for each season and a fifth for the entire year—were the first precipitation maps covering the entire country. A similar set of five maps dealt with temperature.

Coolidge's consent was hardly pivotal: Blodget had been busily mining the charts' "scientific value" for his 536-page *Climatology of the United States*, published in 1857.[26] As in the 1855 *Meteorological Register*, ten maps depicted seasonal and annual temperatures and rainfall, but the underlying data were geographically denser, though less consistent. Sources included the New York academies, the Franklin Institute's correspondents, and a trove of monthly reports solicited by Espy and accumulated at the Surgeon General's Office and the Smithsonian Institution, where its director, Joseph Henry (1797–1878), had hired Espy as a consultant in 1848. Meteorology had caught the fancy of Henry, an eminent physicist—the international unit for inductance is the henry—who foresaw the emerging telegraph network as a way to collect weather data quickly enough to forecast storms. In addition to creating a position for Espy, whose government support was shaky, Henry hired Blodget in 1851 to "reduce" the weather data to more concise tables, amenable to discussion and interpretation.[27] When Henry realized the Smithsonian data were also informing his clerk's own project, he fired Blodget, who retaliated by mentioning the Institution only seven times in his *Climatology* and Henry not at all.[28]

What effect did the richer data set have on Blodget's delineations? To answer this question, I needed to compare the two sets of charts, both missing from the electronic books offered gratis online. The Google Books Project and the Making of America Project, an earlier, less ambitious book digitization endeavor, had independently scanned the 1855 and 1857 volumes, but neither e-book included maps. Scanning bound books is a complex process requiring sophisticated imaging software for removing curvature as well as a quick, reliable process for turning pages. Unfolding and refolding maps tipped into the binding is too slow or cumbersome for either human or computer-controlled page turners, and flagging these omissions for manual intervention later on is not cost-effective—a strong argument, though, for retaining brick-and-mortar libraries with ink-on-paper books. My university library had neither book, but I found a "used" but useful copy of Blodget's *Climatology* through Amazon.com for a surprisingly reasonable $25, plus shipping. The 1855 *Meteorological Register*, with a far smaller print run, was more problematic: several dozen libraries had copies, but interlibrary loans of rare books are difficult to arrange, especially when the borrower wants to unfold fragile charts for scanning. But I found a copy in excellent condition on eBay for $340, which seemed less costly in both time and money than flying to Washington to arrange an expedited scan.

I focused on both books' winter precipitation charts, the only ones relevant to lake-effect snow, and traced their shorelines and rainfall zones onto the paired excerpts in figure 2.3. Tracing was necessary because the faint lines and labels on the original images would not reproduce well at this smaller size, and because the progressively darker gray shadings used to show ever-greater amounts of rain and melted snow are too murky for easy reading. To avoid clutter, I omitted rivers and point symbols and smoothed out minor wrinkles in the shorelines and coastlines but not their salient details, noticeably different on the original charts. Only the chart in the 1857 book included state boundaries, which I retained for a fuller

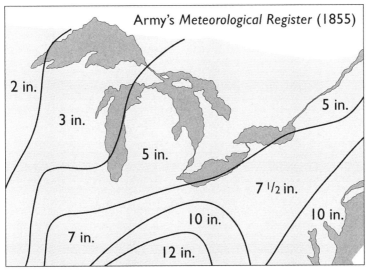

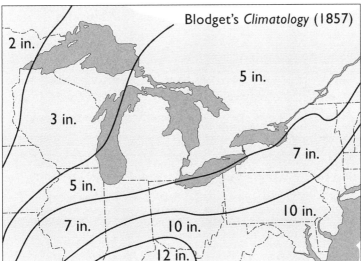

2.3. Generalized excerpts comparing treatment of the Great Lakes region on the winter precipitation charts in the 1855 *Meteorological Register* (above) and Lorin Blodget's 1957 *Climatology of the United States* (below). Numbers represent mean precipitation within the zones. On the upper map a zone with two means (7 and 7½ inches) underscores Blodget's use of cartographic license.

frame of reference. By contrast, the smooth, broad curves of the lines separating precipitation zones required no generalization. I extended coverage
well south of the Great Lakes to show that the rainfall zones are distinctly,
but not radically, different. Blank areas at the top and in the lower-right
corners of both excerpts are no-data zones, too far beyond the observation
network to plot precipitation. The narrower blank areas on the lower map
reflect Blodget's—and the Smithsonian's—denser network of observers,
particularly north of the Canadian border.

Large numerals representing precipitation in inches complete the picture. Although I repositioned the numbers for better visual balance, they
faithfully depict Blodget's identification of precipitation zones by a single
mean value, not a range of values, which had become standard by the
1870s. As explained by identical sentences in both volumes, "The quantities given in the illustration represent the *mean of the areas* in which they
are placed."[29] Blodget's italic *"mean"* is both deliberate and revealing insofar as a statistical mean can obscure substantial variation. Stations within
the zone labeled "5 in.," for instance, should vary around an average of
5 inches, but the zone might well include seasonal averages less than 3
inches or more than 7 inches, the means of adjoining zones. This fudging
exemplifies what map mavens call cartographic license: because his zones
were not defined by explicit intervals, Blodget could draw smooth, approximate, and conveniently imprecise boundaries. His explanation asserts a
cleverness that needs no apology: "The quantities in both cases pass into
each other by gradual transition, and the abrupt distinctions of shading
and of boundary lines are employed only for convenience and clearness
of illustration."

For whatever reason—sparse observations, statistical chicanery, or the
questionable mixing of rain and melted snow—neither map reveals much
of anything about snowbelts or the lake effect. The only hint of enhanced
winter precipitation near the Great Lakes is the boundary between the 5-
and 7-inch zones, which suggests somewhat higher winter precipitation on
the south and eastern sides of Lakes Erie and Ontario. More broadly, the
1857 map's larger data set added little to the dominant pattern of greater
winter precipitation near the coast, ever vulnerable to nor'easters and other
large storms. Coolidge's assurance of exclusive reliance on army data to

the contrary, Blodget might have let the Smithsonian data inform his delineations—in his shoes, I'd find the temptation difficult to resist.

Climatology offered no new insights on lake-effect snow. Not surprisingly, Blodget echoed verbatim the *Meteorological Register*'s remarks on sources of moisture, cold air moving across warm lakes, and orographic precipitation.[30] A page of new material appended to his brief, recycled discussion of snowfall in the United States includes a paragraph on the Great Lakes, apparently based on observer reports. The discussion is wholly descriptive ("The winter snows are often excessive from Buffalo eastward, and they are much more likely to be so than at points west of Lake Erie") and occasionally misleading ("The southern part of the lake district—including the south end of Lake Michigan, the State of Michigan in the latitude of Detroit, and the whole country bordering Lake Erie on the south—is one in which the winter snows melt almost immediately as they fall").[31]

Had Blodget lived near a lake, he might have been as savvy as University of Rochester professor Chester Dewey (1784–1867), who praised *Climatology* in a lengthy 1860 essay in the *North American Review*.[32] Inspired to elaborate, Dewey pointed out that "rain or snow may fall from the condensation of vapor passing into a cooler atmosphere in the vicinity of lakes or large bodies of water." The result "is often noticed in the vicinity of the southern shore of Lakes Ontario and Erie, especially in the fall of snow in small quantities, or to an inch or two in depth, for only a few—from five to ten—miles from the shore." With equal concision Dewey highlighted another lake effect: "the great number of cloudy or hazy nights, especially on or along the south shore of Lake Ontario, protecting a fruit belt of a few miles wide from frost in the spring, so that fruit is rarely killed."

In 1872 the Smithsonian rolled out its own winter precipitation map in a 175-page summary of rain and melted snow measured at 790 stations in Henry's growing network.[33] The report includes three large folded charts, one portraying yearly averages and the others covering the "extreme seasons," summer and winter. The title page names US Coast and Geodetic Survey scientist Charles Schott (1826–1901) as the author but notes that the tables and maps were "discussed under direction of Joseph Henry." Schott said nothing about lake-effect snow and even dismissed the Great

Lakes' contribution to regional rainfall ("Beyond furnishing by their evaporation a supply to the general fund of moisture, the Great Lakes do not
appear to exercise any direct influence. . . . There is even a remarkably
small amount of rain-fall in northern New York, close to Lake Ontario").[34]
As a facsimile excerpt (fig. 2.4) from his winter chart confirms, Schott's isohyets (lines of equal rainfall), drawn to conform to amounts calculated for
individual locations, hold little hint of snowbelts, except perhaps on the
Keweenaw Peninsula, in northern Michigan, where two stations downwind from Lake Superior reported above-average winter precipitation.[35]

Schott and Henry might have seen a stronger link between the Great
Lakes and winter precipitation had they used a more detailed map. In New
York, for instance, their chart shows neither the orographic impact of Tug
Hill nor the comparatively subtle imprint of the Lake Ontario snowbelt,
both readily apparent on the winter precipitation map in the 1893 annual
report of the New York State Meteorological Bureau (fig. 2.5).[36] Although

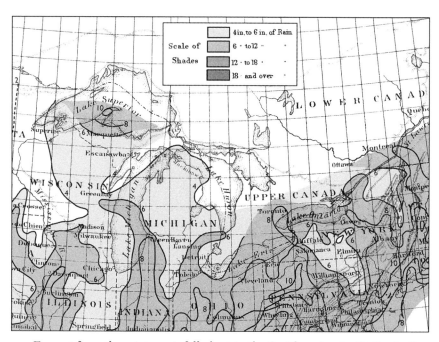

2.4. Excerpt from the winter rainfall chart in the Smithsonian Institution's 1872
report on US precipitation.

the state map is over 10 inches wide and includes county boundaries, rivers, and point symbols for 111 named places, its key features are accurately portrayed by my downsized tracing of its gently curving isohyets and a less cluttered state outline. Isohyets for 10, 12, and 14 inches of precipitation at higher elevations east of Lake Ontario reflect the combined lake-effect/orographic influence, and the contrast between the medium gray along the south and southeast shores of the lake with the lighter shading farther south underscores the influence of proximity. Paradoxically, the more revealing New York map summarizes winter averages for only 80 locations, while its earlier Smithsonian counterpart reflects reports from 132 observers within the state.[37]

Two phenomena with stronger signatures made it difficult for Schott and Henry to recognize the imprint of lake-effect snow on winter precipitation. The dark shading toward the lower-right corner of the New York map indicates the importance of proximity of the Atlantic Ocean, which accounts for relatively high winter precipitation on Long Island, more often as rain than as snow. What's more, higher elevations northwest

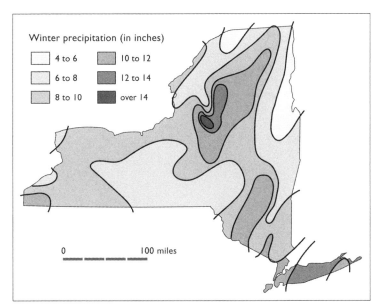

2.5. "Average Precipitation for the Winter" as delineated in the New York Meteorological Bureau's 1893 report.

of New York City in the Catskills, and in the Allegheny Mountains farther west, are generally wetter than surrounding lowlands. These effects,
which the Smithsonian scientists acknowledged, extend well beyond New
York's borders and emerge grudgingly upon close inspection of the tangled
isohyets on their 1872 chart. Even so, on any map based solely on inches
of rain and melted snow, areas to the lee of the lakes do not look markedly
wetter than adjoining areas of similar elevation. It is difficult to discern the
signature of lake-effect snow without measuring snowfall separately and
putting it on its own map.

If there was a defining moment when snow emerged as anything
more than an awkward form of liquid precipitation, it's more easily dated
than explained. In 1814 army surgeons began counting snow days, and in
1870 Congress established the US Army Signal Service, within the War
Department, to issue storm warnings based on thrice-daily weather observations telegraphed to a central forecast office, in Washington. As part of
their duties, observers noted the dates of the season's first and last snows.
In 1872 the Signal Service initiated the *Monthly Weather Review*, which
published scientific articles as well as monthly tabulations of diverse observations, including depth of snow, in inches, at stations reporting the largest snowfalls for the month, and also the range, by state, in the amount of
snow on the ground at the end of each month. Snowfall data were largely
anecdotal until spring 1884, when the Signal Service ordered observers
to report snowfall amounts for twenty-four-hour periods.[38] The network
included not only the 126 stations reporting by telegraph and another 84
reporting by mail but also several hundred additional observers in various
state weather services established in the mid-1880s to disseminate warnings and collect climate data.[39]

Measurement of snow depth remained closely tied to estimates of general precipitation. Observers were to "select a level plane of some extent
and carefully measure the depth of freshly-fallen snow in at least three
places where it has not drifted."[40] If the readings "nearly agree[d]," their
mean was not only reported independently but divided by 10 or 11 to estimate precipitation. If the measurements were too diverse, the observer
(I assume) plunked his ruler or inverted cylinder down in several additional, supposedly representative, locations. Although what qualified as

"nearly" was not defined, errors were assumed to average out around a reliable mean. Because compaction or melting might diminish the observed depth, measurements were taken as soon as the storm ended. Rain changing to snow, or snow to rain, was especially troublesome because an inattentive observer might double count some of the precipitation. Because measuring snowfall was "a very difficult matter [for] rigid rules," Signal Office officials concluded, "each observer must depend largely upon his judgment in each individual case."

Drifting and compaction are hardly the only sources of uncertainty: snowfall can vary enormously from year to year, especially near the Great Lakes. Although year-to-year variations were assumed to average out over time, it was not clear how long a record is required for a reliable small-scale map. The data on hand must have seemed satisfactory to Mark Harrington (1848–1926), who published the first systematic maps of American snowfall in 1894, as part of his eighty-page *Rainfall and Snow of the United States, Compiled to the End of 1891*.[41] An astronomy professor at the University of Michigan with a strong interest in meteorology, Harrington had been appointed chief of the US Weather Bureau in 1891, the year the agency was renamed and transferred to the Department of Agriculture. His interest in snow was handicapped by limited data, only hinted at in the report: none of the tables lists depths or days of snowfall, and little more than half a page of discussion focuses on snow. Some pre-1884 data were apparently useful insofar as he admitted using snowfall records ranging from five to twenty years.[42] His network consisted of fewer than 100 stations: although tables list average monthly precipitation for 103 locations, many are too far south for significant snowfall.

With an extended title that includes "Annual, Seasonal, Monthly, and Other Charts," Harrington's report was printed in two parts: a standard-size volume with text and tables, and an oversize "atlas" printed in color on large (48 × 60 cm; 19 × 24 inches) sheets, which no doubt accounts for its current rarity—according to the electronic catalog WorldCat, only seven libraries own copies. Snow was allotted a single atlas sheet: titled "Snowfall, in inches," it consists of eight small maps, one each for the months October through May. Although none of the maps clearly delineates all of the Great Lakes snowbelts, the October map (fig. 2.6) affords one notable insight: more than five inches of snow to the lee of Lake Superior and more than an inch in a broad area east of Lake Michigan. It's a bit early for lake-effect snow from Lakes Erie and Ontario, and the orographic impact on mountain peaks in the West is conspicuously absent because, as Harrington explained, his maps were "only for snow in the vicinity of the stations."[43]

Today's snowbelts (and most certainly yesterday's) are only faintly apparent in the gently curving 15- and 20-inch isolines on Harrington's December and January maps (fig. 2.7, upper left and right). Although the maps confirm relatively heavy snow in northern Michigan, their isolines reveal nothing exceptional east of Lake Huron or southeast of Lake Erie. The 15-inch line on the December map and the 20-inch line for January indicate more snow immediately east of Lake Ontario than farther south but continue due eastward rather than wrapping around Tug Hill and doubling back; unless the climate had changed markedly since the 1880s (which I

October.

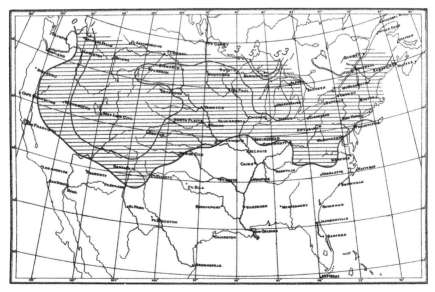

2.6. Harrington's map for October shows an average of more than five inches of snow near Marquette, Michigan. The eight-map Snowfall sheet was printed in orange and dark blue inks.

doubt), an accurate map would show less snow around the northeast end of the lake. What's more, reliable data for at least a few Canadian stations would have shown noticeably more snow in Buffalo than in Toronto.

More precise isolines with an interval less than five inches might have been more revealing. Harrington, who no doubt pored over the data as carefully as he looked at the maps, got it right in observing that "the area of deep snows for southern Michigan is from the middle of the west coast, in the vicinity of Manistee, nearly straight across the peninsula," and that "the area of deepest snowfall in New York is to the eastward of Lake Ontario, and, to some degree, to the southward, in the immediate vicinity of the lake." It is not readily apparent in his atlas, though.

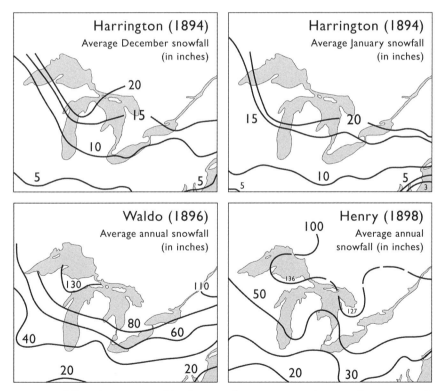

2.7. Snowfall in the vicinity of the Great Lakes, as portrayed on Mark Harrington's 1894 snowfall maps for December (*upper left*) and January (*upper right*), Frank Waldo's 1896 map of average annual snowfall (*lower left*), and Alfred Henry's 1898 map of average annual snowfall (*lower right*).

Despite an ambiguous portrayal of Lakes Erie and Ontario, the maps reinforced a growing perception of lake snow and its key ingredients. In noting that "the area for deepest snow in the United States not mountainous is along the south shore of Lake Superior, from Marquette eastward"— a fact blatantly obvious on most of the monthly maps—Harrington opined that "this would quite agree with the suggested influence of the Lakes, in that the air passing over Lake Superior comes largely from the northwest, and by the time it reaches the coast in question has already received a surcharge of vapor chilled by the surface of the lake." However hazy the cartographic evidence, lake-effect snow was now a hypothesis worth further study.

Could Harrington have discerned the Great Lakes snowbelts by mapping average annual snowfall? Not necessarily. In 1896, Frank Waldo (1857–1920) published *Elementary Meteorology for High Schools and Colleges*, which included the first map of average annual snowfall for the United States (fig. 2.7, lower left). Waldo was a Harvard graduate and former Signal Service employee whose résumé included cattle ranching, textbook writing, and lecturing at the Evelyn College for Young Women, in Princeton. It's not clear where he got his data. He didn't credit the Weather Bureau or its chief, but he might have added up the snow depths on Harrington's eight monthly maps, which seems likely because he not only cited the seven-year period 1884–91 but also noted that his map shows snowfall only for "the low lands and plains in which the larger towns are located."[44] As with Harrington's maps, his chart captured the lake effect only for Michigan's Upper Peninsula. Unlike Harrington, he drew no connection between the Great Lakes and locally heavy snow.

Two years later, the *Monthly Weather Review* included a map of average annual snowfall by Alfred Judson Henry (1858–1931), chief of the Weather Bureau's Division of Records and Meteorological Data and no relation to the Smithsonian's Joseph Henry. An accompanying table lists average snowfall for 159 places in the United States with at least a "trace" of snow and another 29 in Canada.[45] While there's no doubt where Henry got his data, the period of record ranged from three to eleven years, and surely included measurements as recent as 1895. A short description of the map and table ignores general trends and possible causes.

Henry's map (fig. 2.7, lower right) is more informative than Waldo's. Dashed portions of his 100-inch snowfall line reflect a wary treatment of thinly settled parts of Canada. Eager to show noteworthy anomalies, he included depth figures for "special localities" like the south shore of Georgian Bay (127 inches) and the south shore of Lake Superior (136 inches). Lacking isolines between 50 and 100 inches, the map ignores lake snow east of Lake Ontario and only faintly hints of a Lake Erie snowbelt between Cleveland and Buffalo. The *Review* had published monthly snowfall maps since 1890, but this map was the first averaged full-year summary.[46] It did not become a regular feature.

Nor did a map of snowfall win a place in Henry's massive *Climatology of the United States*, published in 1906. Lack of data was not a problem: the book's 1,012 pages include tables listing monthly, seasonal, and yearly averages for depth of snow at more than 600 locations—volunteer observers using standardized instruments and recruited by the state weather services provided broad, presumably reliable coverage of the nation's climate by filling gaps between Weather Bureau forecast offices.[47] The network was evidently adequate for large charts for the first killing frost in autumn, the last killing frost in spring, and the average annual number of thunderstorm days, but not snowfall, covered cartographically in a watered-down chart of normal annual precipitation.

Henry devoted only a page and half of text to snowfall, but his discussion seems well informed. Either he had made a snowfall map that he didn't include or he had a good mental map, complete with 5- and 50-inch lines of annual snowfall, the westward trends of which he described in moderate detail.[48] ("The line of 50 inches annual snowfall enters the United States on the Massachusetts coast, passes thence a little south of west to the upper Ohio Valley, . . ."). Unlike Harrington's tentative, somewhat rambling comment twelve years earlier, his one-sentence assessment of lake-effect snow was confident and concise: "As the winds over the Great Lakes in winter are mostly west to northwest, the snowfall on the leeward side of the lakes is much heavier than on the windward side."

The growing accumulation of snowfall data was an irresistible invitation to Charles Franklin Brooks (1891–1958), who received a bachelor's degree from Harvard University in 1911 and stayed on for graduate work with the eminent climatologist Robert DeCourcy Ward (1867–1931). During his first year as a graduate student, Brooks wrote a paper on snowfall in the United States, which so impressed Ward that he sent it to London, where it was read with his endorsement in January 1913, at a meeting of the Royal Meteorological Society, and published that April in the society's *Quarterly Journal*.[49] It was a short paper in which Brooks moved quickly from the relative neglect of snowfall "in the early days" of the Signal Service to the initiation of standardized measurements in 1884, to a table listing the numbers of stations reporting monthly snowfall by year, from 1884–85 through 1909–10, to discussion of the maps of Harrington, Waldo, and Henry as well as his own national map of average annual snowfall, based mostly on data from July 1895 through June 1910 for 159 stations with a continuous fifteen-year record.

A revealing snowfall map, Brooks believed, demanded cautious expediency. "To give some idea of the extent of the snowfall of the country" and "considerably reduce the probability of error," he supplemented the 159-station network with three additional sets of "necessarily unsatisfactory data" for progressively shorter periods, the last covering only one to three years. While terrain maps provided useful guidance, particularly in the mountainous West, "in drawing the lines of equal depth of snowfall, the writer used his own discretion pretty freely"—an "exercise of personal judgment" defensible until "more, and better snowfall data, are available." Though wary of the "difficulties met with in the construction of the present chart," he was confident it "represent[ed] the actual conditions of snowfall a good deal more accurately than the charts which have preceded it."

A generally complete picture of Great Lakes snowbelts rewarded Brooks's optimism. In contrast to snowfall maps published in the 1890s (fig. 2.7), his 100-inch isolines (fig. 2.8) reveal well-above-average snowfall not only south of Lake Superior but also to the lee of Lakes Erie and Ontario. The most accurate of his predecessors was Alfred Henry, whose 50-inch isoline portrays locally heavy snowfall along the northern two-thirds of the Lake Michigan's eastern shore: a zone better captured by Brooks's 50-inch isoline, which extends southward into northern Indiana, as on Val Eichenlaub's map (fig. 1.1). The most obvious omissions are the snowbelts in Canada east of Lakes Michigan and Huron. Whether wary of Canadian data or the extra work entailed, Brooks did not draw snowfall lines north of the border. This omission was pointed out by the first of nine discussants who volunteered opinions after the paper was read aloud at the RMS, but most offered praise and encouragement as well as advice. Several would have preferred a longer paper, with more analysis and interpretation. Brooks said little about what his map showed, but he included the Great Lakes in a short list of "effects" deemed "very apparent."

Fortunate to have found a workable dissertation topic early in his program, Brooks wisely narrowed the geographic scope of his doctoral research to the eastern United States. The data were comparatively complete east of the Mississippi River, and including the West would have been pointless as well as burdensome because of its rugged terrain and complex climatology. With electronic computing a half century in the future, the research entailed an enormous amount of hand calculation. Brooks confessed an appreciation of then-current technology in noting "the data were reduced with the aid of an adding machine."[50]

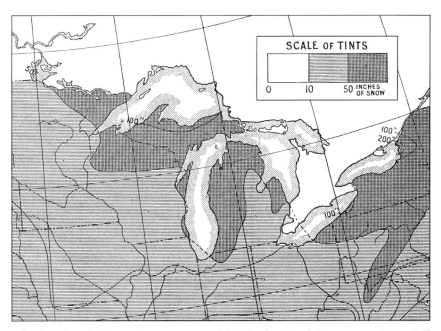

2.8. The Great Lakes region as portrayed in Charles Brooks's 1913 map "Snowfall of the United States."

The core of the study is an "atlas" of fifteen large charts, 17 inches high by 14 inches wide, bound separately from the rest of the dissertation. Chart I is merely the eastern half of the US Geological Survey's 1911 Relief Map of the United States, included without annotation as a reference; its nine color-coded elevation categories (only six occurring in the East) differentiate lake plains, coastal plains, and river valleys from mountains and their foothills. Brooks underscored the importance of elevation by plotting lines of equal snow depth on the eastern portion of the companion USGS Contour Map, printed at the same scale (1:7,000,000), with elevation contours at a 1,500-foot interval. Ten copies of the less visually busy Contour Map provided the framework for an annual average snowfall map and individual monthly maps for September through May. Brooks acknowledged the significance of wind direction with four other monthly maps, for December through March, on which eight-pointed wind roses describe "snow-bearing winds" for fifty-eight suitably separated sites. Thirty-four additional maps and statistical charts occupy twenty-seven of his dissertation's ninety-one letter-size pages and attest to early-twentieth-century climatology's heavy reliance on data graphics.

Compilation was a two-stage process. Before plotting his thick isolines on the USGS base maps, Brooks worked up "large scratch maps of the individual states." He based his snowfall lines on stations with eighteen years of comparable data, from fall 1895 through spring 1913, "except where stations with such records were too far apart."[51] In filling the gaps, he could usually rely on records fourteen to seventeen years long; the principal exceptions were the Appalachians and "sparsely populated" portions of Michigan, Minnesota, and Wisconsin. Although the records for some stations went back more than eighteen years, he ignored the earlier years, citing the importance of homogeneity "to bring out the true geographic relations of such a variable climatic factor as snowfall."[52] Even so, he was roundly skeptical that data collected "only once a day or once a snowstorm" were not affected by "drifting, wind and rain-packing, melting and evaporation" as well as "further errors . . . in the printing, tabulation, and reduction of the data."[53] Like many scientists, yesterday and now, he believed that "on the final maps, the effects of such errors seem to be practically eliminated through the use of extensive data."

However flawed his data, Brooks's maps were noticeably more detailed than earlier attempts and more closely aligned with emerging recipes for lake-effect snow. In contrast to the delineations on his 1913 map—200 inches of snow immediately east of Lake Ontario (fig. 2.8) is surely a fluke—his new average annual isolines for the Great Lake region, which I retraced in figure 2.9, afford a nuanced portrayal of heavy snow in the highlands immediately east of Buffalo and also on Tug Hill, east of Lake Ontario. Although the Adirondacks registered the highest snowfall in the East, with over 130 inches near an isolated hamlet named Number Four, Tug Hill scored nearly as high, with a small, closed 120-inch isoline near Adams, New York, south of Watertown. Collectively the 50-, 70-, and 100-inch isolines reflect all the Great Lakes snowbelts on US soil. The monthly snowfall and winds charts are similarly revealing.

Brooks's ten-page discussion of Great Lakes snowfall is a clear, concise, and confident description of the lake effect. "In winter," he noted, "moist cyclonic winds, blowing from a comparatively warm water surface

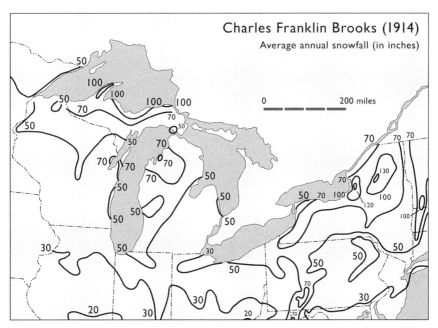

2.9. The Great Lakes region as described by the average annual snowfall map in Brooks's 1914 dissertation.

onto cold land, cause increased precipitation . . . chiefly by forced ascent of the wind due to increased friction and to topography [resulting in] heavy snowfall on leeward shores," which receive "fairly regular west-wind snowfall plus the irregular cyclonic snowfall" of passing storms.[54] Snowfall is less variable along leeward shores than in windward locations, which receive almost all of their yearly snow from a few large cyclonic storms. A page-size map of snow days confirmed the importance of persistent winds from the west and northwest: unlike most of the east, the snowbelts typically experience more than fifty snow days a year. And even though lake snow is less common in late winter, when much of the water surface is frozen, "drifting snow off the lake" can augment snowfall along the shore.[55]

Like many other new PhDs, Brooks published a few articles from his dissertation and moved on to other projects.[56] He helped found the American Meteorological Society in 1919, held teaching jobs at Yale and Clark, and joined the Harvard faculty in 1931. Between 1923 and 1927 he wrote a syndicated daily newspaper column, *Why the Weather?* but only two of several hundred installments—"Snowsqualls of the Great Lakes," for November 2, 1923, and "Snowfall about the Great Lakes," for the following December 17—address lake-effect snow.[57] The latter offers an impressively concise descriptive interpretation: "There is still much open water to moisten the passing air and there are frequent cold west and northwest winds to drive under and elevate the air over the Lakes. Even the small obstruction offered by the low eastern shores does its part in forcing the strong Lake winds to rise, thereby cooling them further and increasing the snowfall."

In 1936 Brooks collaborated with Canadian climatologist A. J. Connor on *Climatic Maps of North America*, a folio collection of twenty-six large (22 × 17 inch) maps that includes a chart for average annual snowfall. As the excerpt in figure 2.10 confirms, the map covers Canadian territory around the Great Lakes, and its snowfall lines depict, or at least suggest, snowbelts east of Lakes Superior and Huron as well as to the lee of Lakes Erie and Ontario. The authors claimed their chart "shows exceptional detail" because they used data for 2,600 stations in the United States and another 800 in Canada, but the lack of a 100-inch snowfall line produced a vague portrait of lake-effect snow, especially around Lake Michigan.[58]

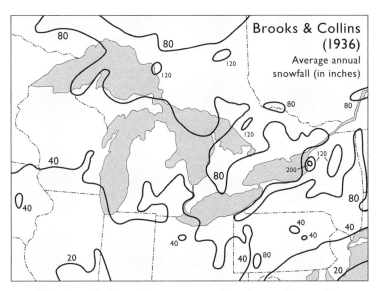

2.10. The Great Lakes region as portrayed on the 1936 map of average annual snowfall by Charles F. Brooks and A. J. Connor, redrawn at a smaller scale.

Excessive generalization—note, in particular, the large island missing between Georgian Bay and Lake Huron—might reflect the chart's preparation for the North American volume of the ambitious *Köppen-Geiger Handbuch der Klimatologie*, a collaborative international encyclopedia begun in 1927 but never completed.[59] Although the project's goal of comprehensive worldwide coverage made it impossible to ignore snow as a distinct and significant weather element, the *Handbuch*'s standardized format undermined treatment of the Great Lakes region, which was poorly served by cramming the entire continent onto a single small-scale map sheet and minimizing graphic congestion in the Rockies with a sparse set of isolines for the entire map.

While Brooks was publishing the results of his dissertation, a team of specialists at the Department of Agriculture was at work on the massive *Atlas of American Agriculture*, the brainchild of Oliver Edwin Baker (1883–1949), an economic geographer with broad interests in agriculture and population and a varied career of university teaching and government service.[60] A complex project addressing farmers and their families, rural land use, natural vegetation, and the physical requirements of agriculture, the *Atlas* was issued in multiple sections between 1918 and 1936, its lengthy gestation a reflection of cost, complexity, and World War I.

Snow was covered in the Precipitation and Humidity section, which was ready in 1917 but delayed until 1922 because of wartime priorities and a paper shortage. Compiled by climatologist Joseph Burton Kincer (1874–1954), the Precipitation folio relied on records from 1,600 stations for the twenty-year period 1895 to 1914, and included 69 maps, large and small. Robert DeCourcy Ward, who reviewed an advance copy in the *Monthly Weather Review*, called it "a very important advance in the accurate charting and discussion of many of the essential features of the climates of the United States."[61]

One the section's smaller maps shows the general pattern of snow days, defined as the average annual number of days with 0.01 inch or more of melted snow—a prudent specification that excludes days with just a few flurries and recognizes regional differences in the snow-to-liquid ratio. Reproduced in black-and-white in figure 2.11, the chart shows only three parts of the country with more than 60 snow days: an orographic snowbelt stretching from northwest Wyoming into Montana, an area running westward from Michigan's Upper Peninsula into North Dakota, and a zone stretching from the eastern end of Lake Erie along the southern and eastern shore of Lake Ontario into the Adirondacks—the latter two zones partly confirm the Great Lakes' reputation for persistent if not intense snowfall. Another small chart further on reveals a yearly average of more than 160 cloudy days along the south shore of Michigan as well as east of Lakes Erie and Ontario, a lee-of-the-lake effect shared with much of Washington State, downwind from the world's largest water body, the Pacific Ocean.[62]

Additional charts show the average date of the first snowfall, the average annual number of days with snow on the ground, and the average annual snowfall "in inches, unmelted." Reproduced at a larger, more generous scale than its *Handbuch* counterpart, Kincer's snowfall map, with a comparatively detailed array of isolines—20, 30, 40, 50, 60, 80, 100, 120, and 150 inches in the excerpt in upper part of figure 2.12—affords a

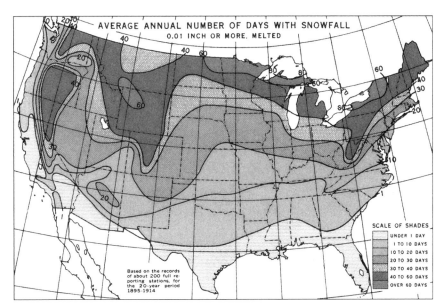

2.11. Snow-days map in the Precipitation and Humidity section of the *Atlas of American Agriculture*, issued in 1922.

clearer depiction of Great Lakes snowfall. Areas with significant lake snow or lake-enhanced orographic snow are readily apparent (at least inside the US border) to the lee of all the Great Lakes, except Lake Huron, a prime contributor to snowbelts in Ontario. Despite this obvious relationship, Kincer was steadfastly descriptive, refusing to suggest an explanation in his accompanying thousand-word discussion and not mentioning the Great Lakes by name, individually or collectively.[63] We're told, though, that more than 400 inches of snow have been recorded on the west side of the Sierra Nevada and the Cascades, and that snowfall in the Great Plains ranges from an inch in central Texas to about 20 inches in northern Kansas. Lots of place names and numbers, but no reference to lake snow.

Kincer was no more informative two decades later, when he wrote the chapter "Climate and Weather Data for the United States" for the 1,260-page *Climate and Man*, published as the 1941 Yearbook of Agriculture.[64] A smaller map, run lengthwise on a full page in the book-format Yearbook, affords a less spatially detailed picture of average annual

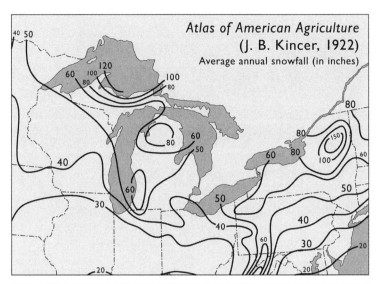

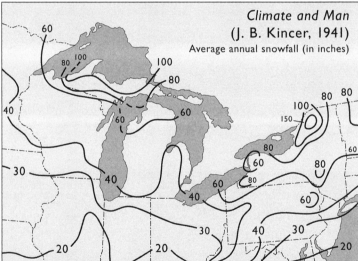

2.12. Average annual snowfall in the Great Lakes region, as portrayed on Joseph Kincer's maps for the *Atlas of American Agriculture* (upper) and *Climate and Man*, the Yearbook of Agriculture for 1941 (lower). The forty-year period (1899–1938) for the lower map partly overlaps the twenty-year period (1895–1914) for the upper map.

snowfall (fig. 2.12, lower). While the isolines are similar in value to those on his 1922 snowfall chart, their configuration seems more generalized—understandable, though, because of the smaller scale. Oddly, the zone of peak snowfall in the Adirondacks has shifted to the northwest, and there's no hint of a snowbelt along the southeast shore of Lake Michigan. These differences at least partly reflect a longer period of record, 1899–1938, but there's no comparison, discussion of trends, or even a litany of regional differences.

Stephen Sargent Visher (1887–1967), who replicated the isolines of Kincer's snowfall and snow-days maps in his 1954 *Climatic Atlas of the United States*, was not shy about interpreting maps.[65] "The influence of the Great Lakes is discernible on most of the climatic maps here being studied," he opined, and invited readers to examine his maps of average, minimum, and maximum temperatures for January, July, and the full year as well as those for the frost-free period, seasonal precipitation, and snow. Visher affirmed the existence of snowbelts in noting that the lake's "effect upon snowfall is appreciable; on average, they increase it about 20 inches, but in Upper Michigan and just east of Lake Ontario, the increase is more than 40 inches."

In 1970, when the US Geological Survey published the long-awaited *National Atlas of the United States*, the period of record for the average annual snowfall map advanced to 1931–52. Although the *National Atlas* had been underway since 1954, when the National Academy of Sciences set up a planning committee, the throes of research and Congressional fund-raising cannot account for the eighteen-year lag: the map of mean number of snow days, which shared a page in the *Atlas* with the map of snowfall, covered a longer, more recent period, 1931–64.[66] According to the fine print, the snowfall map had been "adapted" from a larger-scale map by the Environmental Data Service (EDS), a part of the new Environmental Science Services Administration (ESSA), formed in 1965.[67] Apparently lacking an off-the-shelf snow-days map, the EDS prepared one for the *Atlas* and used its most recent data.

Although constrained by map scale and unique isolines, both *National Atlas* maps not only confirm lee-of-the-lake snowbelts along the American shores of the Great Lakes but suggest different levels of lake-snow intensity. I highlighted the two highest intervals in their excerpts (fig. 2.13), which underscore the added orographic effects at Tug Hill, east of Lake Ontario, as well as the rugged terrain south of Lake Superior. The 64-inch snowfall line (fig. 2.13, upper) completes the picture by encompassing zones of the less intense lake snow east of Lake Michigan, south of Lake Ontario, and along the far eastern end of Lake Erie—second-tier snowbelts not readily differentiated from highlands receiving substantial helpings of Atlantic snow by an isoline that meanders eastward to encircle the Catskills and the Adirondacks.

Similarly, on the snow-days map (fig. 2.13, lower) round-number isolines extend well beyond the recognized snowbelts, most noticeable where the Appalachians trend south from Pennsylvania into West Virginia. Even so, the curvature of the snow-days isolines more clearly reveals the influence of the Great Lakes than Kincer's 1922 snow-days map (fig. 2.11), for which any day with only a 1/100 inch of melted snow counted as a snow day—according to the 10:1 rule; that's only a tenth of an inch of actual snow, well below the more demanding and meaningful one-inch threshold for the *National Atlas* map.

To better understand how map scale and isoline values can affect a snowfall map it's useful to look at the 1:10,000,000 EDS map on which the *National Atlas* staff based their smaller 1:17,000,000 snowfall map. Quick refresher: 1:10 million means that an inch on the map represents 10 million inches on the ground. (The 1:17 million map is said to have a "smaller" scale because 1/17 is a smaller fraction than 1/10.) What's important is that the 1:17 million *National Atlas* snowfall map must cram the same territory into a much smaller space than its EDS counterpart, which occupied a full page in the *Climatic Atlas of the United States*, published in 1968. And as my Great Lakes excerpt (fig. 2.14, upper) demonstrates, the EDS map did indeed have more isolines, which in a somewhat different way reflects snowbelts south of Lake Superior and east of Lakes Michigan, Erie, and Ontario.

Why do the two snowfall maps look so different? Much of the answer lies in the formulas underlying their isolines. To avoid appearing arbitrary,

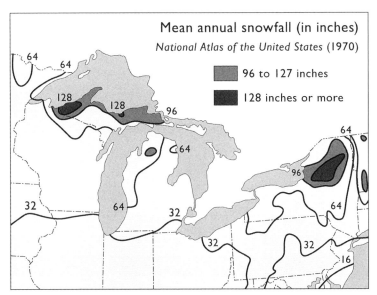

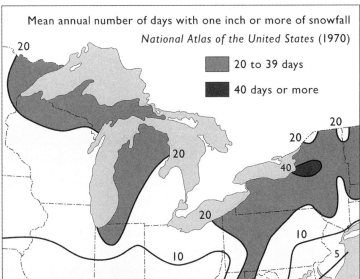

2.13. Snowfall (*upper*) and snow days (*lower*) in the Great Lakes region, as portrayed on the Snowfall sheet in the *National Atlas of the United States*. Note that the snow-days map is based on a one-inch threshold, which ignores differences in the snow-to-liquid ratio.

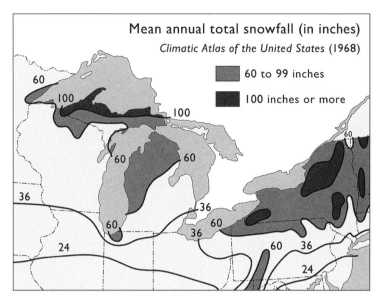

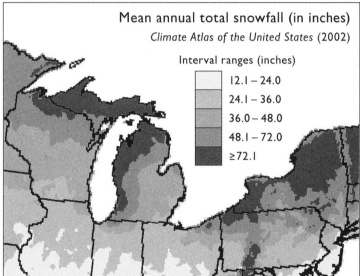

2.14. Snowfall in the Great Lakes region, as portrayed in the 1968 *Climatic Atlas of the United States* (*upper*) and the 2002 *Climate Atlas of the United States* (*lower*).

a cartographer typically adopts a rule like the powers-of-two strategy that populated the *National Atlas* map with isolines representing 8, 16, 32, 64, and 128 inches of snow but added a 96-inch line for places like the Adirondacks and the south shore of Lake Superior.[68] Whoever compiled the snowfall map for the *Climatic Atlas* chose a different rule, based on fractions and multiples of 12, which fixed the isolines at 1, 2, 4, 6, 12, 24, 36, and 60 inches, with a conventionally rounded 100-inch line thrown in for heavy-snow areas.[69] With different sets of isolines, the maps offer different takes on the complexities of average annual snowfall. Less obvious is the strategy for interpolating intermediate isolines: although the 32-inch *National Atlas* line (fig. 2.13, upper) seems a smoother, slightly more southerly version of the 36-inch EDS line (fig. 2.14, upper) from which it was "adapted," the 64-inch *National Atlas* line probably reflects a peek at the underlying numbers, an elevation map, or both—or perhaps even a subconscious effort to mimic snowbands.

A final map excerpt demonstrates that mapping snowfall with isolines requires further generalization if the cartographer wants to avoid the clutter of lines that loop back on them themselves or crowd their neighbors. A skeletal set of isolines, smoothed and displaced to avoid overlap, is unavoidable when visually distinct lines represent specific amounts of snowfall. This approach exemplifies vector graphics, composed of lines plotted in pen and ink or stored electronically as lists of coordinates. By contrast, raster graphics, based on numbers organized in rows and columns like the pixels in a digital photograph, avoid graphic congestion with an array of small squares, each colored to represent a specific range of values, such as 24.1 to 36.0 inches of snow. As demonstrated by the Great Lakes excerpt (fig. 2.14, lower) from the *Climate Atlas of the United States*, published in 2002 by the National Climatic Data Center and offered online, a raster map affords a more fine-grained cartography of snowfall—except where ugly pixilated boundaries or shorelines overwrite the color-coded cells. No map is perfect.

Does the NCDC map present a more precise picture of Great Lakes snowbelts? In a word, no. Its added spatial detail is an improvement, to be sure, but 72 inches is much too low a threshold for the top category, which fails to differentiate significant snow from the really heavy snow found south of Lake Superior and east of Lake Ontario, on Tug Hill. In my own experience, lumping Syracuse, with 119 inches of snow in an average year, with Boonville, New York, where the yearly average exceeds 200 inches, is misleading. Because the new map's uppermost category is too broad, its snowfall lines cannot reflect the existence and extent of lake-effect snowbelts any better than the overly generalized snowfall lines in the *National Atlas*.

To appreciate the importance of spatial detail it's useful to look at John Henry Thompson's *Geography of New York State*, a collection of essays by various specialists published in 1966.[70] Among the folded maps in a pocket glued to the inside back cover is Mean Seasonal Snowfall, which opens to a sheet 21 inches wide by 16¼ inches tall. The excerpt in figure 2.15, reduced here from a width of five inches, is far more detailed than any of snowfall maps examined earlier. The map was prepared in 1960 by Bob Muller, then a geography graduate student at Syracuse University. Two- or three-digit numbers show average annual snowfall, in inches, for stations represented by large black dots. The numbers help readers assign values to the isolines, plotted at a 20-inch interval where the average snowfall is less than 100 inches and at a 30-inch interval elsewhere. A sequence of ever darker graytones, printed in blue on the original, represent progressively heavier snowfalls: under 60 inches for the lightest tint, 60–100 and 100–160 inches for the next two shadings, and over 160 inches for the darkest gray. Although the map confirms the importance of west-to-east winds and proximity to the lake, the Lake Ontario snowbelt is hardly homogeneous. Tug Hill and lesser highlands partly visible along the right and bottom edges of the excerpt accentuate the role of elevated terrain in wringing additional moisture out of lake-induced snowstorms.

Muller's intricately crafted isolines are no accident. As a geographer who recognized that a snowfall map, like any cartographic drawing, is inherently subjective, he began by plotting snowfall data on a contour map, which helped him incorporate the plausible effects of topography on exposure to diverse types of snow-bearing winds, including the passage of warm fronts and cold fronts, orographic fallout from polar air masses, and lake squalls. The result was an interpretative map that more accurately

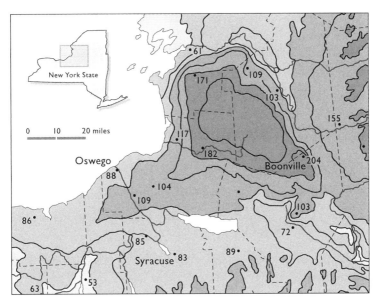

2.15. Oswego County and Tug Hill as portrayed on the map of Mean Seasonal Snowfall compiled by Robert A. Muller. On the original map the area in this excerpt is five inches wide.

reflects the pattern of snowfall than earlier, small-scale charts, which largely ignored local topography. In 1966, in a short article in *Weatherwise* titled "Snowbelts of the Great Lakes," he unveiled a more generalized version of his New York map extending from Maine to Minnesota. "The most outstanding features of the map," he noted, "are the snowbelts associated with the frequently recurring lake squalls coming off the Great Lakes."[71] Muller identified six snowbelts, roughly represented by the 100-inch snowfall line, but he did not delineate their boundaries, as Val Eichenlaub did four years later in the more prestigious *Bulletin of the American Meteorological Society*.[72]

A new genre of snowfall map emerged in the final decades of the twentieth century. Climatology had leaped from tabulating and mapping averages, extremes, and monthly time series to assessing risk and mapping probabilities, and this transition led to maps showing the probability of a white Christmas or the likelihood of six or more inches of snow during the first half of February. An amalgam of computer modeling and statistical analysis, this approach became a standard risk-management tool for studying natural hazards of all types, including earthquakes, flooding, and severe windstorms. A risk map typically links an event of a particular magnitude to a specific time interval (sometimes called the *recurrence interval*) and an *exceedance probability*. For instance, a seismic planning map used to help engineers and architects design earthquake-resistant bridges and structures might describe variation throughout a region in the ground acceleration (represented in g units as a proportion of the acceleration of gravity) with a 2 percent chance of being equaled or exceeded over a fifty-year period.[73] Similarly, a map showing the percentage of days with four or more inches of snowfall between November 1 and 15 would help a local highway superintendent plan for the forthcoming plowing season as well as inform an entrepreneur eager to develop a new ski resort.

In 1993 the Northeast Regional Climate Center, a NOAA affiliate at Cornell University, published an atlas with 193 probability-based snowfall, snow-day, and snow-cover maps for an area stretching from West Virginia to Maine.[74] The maps cover quarter months, half months, whole months, and the full season; their themes address snowfall for the period, maximum snow depth for the season, and the percentage of days with a particular amount of snowfall or snow cover; and most of them include southeastern Canada. Because almost all of the maps are useful in assessing both the impact and the seasonal progression of lake-effect snow, selecting an example was difficult. For the excerpt in figure 2.16 I chose the NRCC's 95th-percentile snowfall map for December, a month that exemplifies the lake effect. Isolines not only describe snowbelts east of Lakes Huron, Erie, and Ontario but also pinpoint areas where the December snowfall should not exceed 70 inches nineteen years out of twenty, on average. In general, an exceptionally snowy December near the Great Lakes will involve markedly more snow than an unusual December farther east or south, near the coast.

Perhaps the most intriguing cartographic portrait of Great Lakes snowbelts appeared in 1996, in an article titled "Impacts of the Great Lakes on Regional Climate Conditions."[75] Curious about the lakes' contribution to

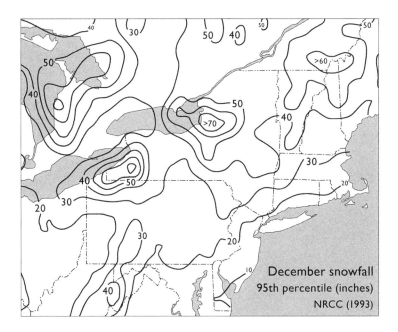

2.16. December snowfall expected to be exceeded one year out of twenty in the eastern Great Lakes region, as shown on the 95th-percentile December snowfall map in the Northeast Regional Climate Center's snowfall atlas.

annual and seasonal patterns of temperature, precipitation, cloud cover, wind speed, and humidity, Robert Scott and Floyd Huff, meteorologists at the Illinois State Water Survey, made two sets of maps, one for all climate stations in the region and another that ignored stations within eighty kilometers (fifty miles) of any of the Great Lakes. They then subtracted each "no-lake-effect" pattern from its corresponding all-stations map to produce a map of "lake-induced changes." Figure 2.17, which shows lake-induced winter precipitation, in millimeters, reflects all six of Eichenlaub's and Muller's snowbelts. Although the zone east of Lake Michigan is less robust than the other five, the orographic effect of the Porcupine Mountains on Michigan's Upper Peninsula is particularly prominent. While some of the precipitation fell as rain, not snow, and even though an eighty-kilometer

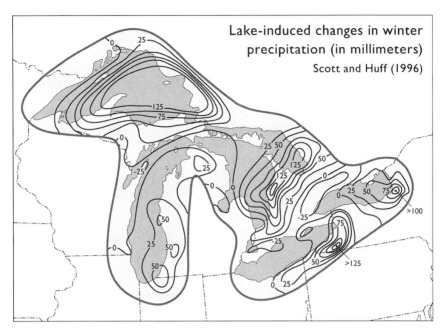

2.17. Lake-induced winter precipitation, in millimeters, within 80 kilometers of the Great Lakes. The thick black line is the outer boundary of the 80 km buffer. Negative numbers indicate areas in which presence of the lakes might have reduced winter precipitation.

buffer cannot isolate the lake effect entirely, this view of lake-induced winter precipitation further confirms the significance of lake-effect snow.

The maps in this chapter, which document the cartographic discovery of the Great Lakes snowbelts, reflect the importance of a suitably dense network of climate stations with standardized and reliable instruments as well as the value of mapping in making atmospheric phenomena legible by linking separated measurements. The next chapter moves from the climatologist's maps of averages based on decades of data to the meteorologist's cartographic snapshots of individual storms, indispensable for understanding lake-effect snow and forecasting storms. Collectively, the two chapters show that as meteorology evolved into a science driven by both data and theory, atmospheric scientists recognized lake-effect snow as a distinctive though often troublesome form of winter precipitation.

3 Prediction

The federal weather service, organized to warn of synoptic storms and other large weather systems, was slow to ponder, much less forecast, lake-effect snow. Even so, the Great Lakes' contribution to winter weather incited a curiosity that led to recognition that lake-effect snow was different from ordinary snow. But unlike William Redfield and James Pollard Espy, whose controversial storm theories and heated rhetoric anticipated the post–Civil War weather bureaucracy, the government forecasters whose short but insightful writings inspired a fuller understanding of lake snow are recognized for other accomplishments, if at all.

Few present-day meteorologists are likely to have heard of Charles L. Mitchell (1883–1970) or Wilfred P. Day (1890–1982), rock stars of weather prediction in the 1930s and 1940s, when they worked at the US Weather Bureau's district forecast office in Washington, DC. Mitchell, who retired in 1950 after forty-six years of service, was known for his exceptionally accurate extended forecasts, one of which contributed to the successful Allied invasion of North Africa in 1942.[1] Day, who received a medal for meritorious service in 1954, was dubbed the "Dean of District Forecasters" when he stepped down in 1957, after forty-four years with the Bureau.[2] Focused on operational meteorology, neither had much time for writing research papers. Even so, Day authored eighty-three notes, most less than a page in length, published between 1920 and 1929 in the *Monthly Weather Review*, to which Mitchell contributed eight somewhat longer articles between 1920 and 1933.

Despite faint bibliographic footprints, Mitchell and Day provided what might be the earliest, and perhaps the most concise, scientific explanation of lake-effect snow. Their two-page article "Snow Flurries along the Eastern Shore of Lake Michigan" appeared in the September 1921 issue of the *Monthly Weather Review*. Mitchell, listed as the author, credited Day for the piece's sole illustration.[3] They had been colleagues at the district forecast office in Chicago for several years before the Weather Bureau, around 1920, transferred them to its lead forecast office, in Washington, DC, where Day's father was the chief climatologist. According to the *Official Register of the United States*, which lists annual salaries and job titles, Mitchell earned $2,520 as a meteorologist, while Day received $1,440 as an "observer."[4] In the decades to follow, Mitchell was always a pay grade or two ahead of Day.

What Day contributed to the collaboration was several years at the Weather Bureau office in Ludington, Michigan, his first posting, and an aptitude for sketching, honed off-hours at the Chicago Academy of Fine Arts. Ludington is midway along the lake's eastern shoreline, and conversations about Day's experiences might have roused Mitchell's curiosity about lake snow. "How far back over Lake Michigan [did] these flurries extend," he wondered, and what caused them? For an answer he wrote officials of the Pere Marquette Railway, which operated car ferries from Ludington across to Milwaukee and Manitowoc, Wisconsin. The three ferry masters who replied recalled seeing steam or fog rising as far back as the lake's western shore, which typically meant snow east of the lake, particularly in early winter, when the water surface, at around 40°F in early December, was much warmer than frigid air from the west and north. Mitchell reasoned that the lake was covered by "a layer of warmer air" that was "necessarily quite shallow along the western shore" but thicker "toward the east," where "convectional currents and turbulence set in, manifesting themselves in . . . clouds farther out in the lake and . . . snow flurries where convection and turbulence are sufficient to produce it."[5] Although lake snow could be intense, Chicago forecasters used "snow flurries" to distinguish it from conventional snowstorms associated with centers of low pressure.

Day summarized the process with the short, wide, and insightfully concise diagram in figure 3.1. With appropriate vertical exaggeration, it reads from left to right, like a sentence, to explain a process that evolves from west to east. The tale begins with rising water vapor as cold winds overrun the lake's relatively warm water. About halfway out the rising vapor triggers convection, which then carries the condensing moisture to still higher levels, creating turbulence, tall clouds, and "snow flurries" over land. As a cross-sectional view that's also a time line, Day's graphic narrative nicely complements today's overhead images, captured by satellite or constructed with Doppler radar, in which the two-dimensional form of a typical snowband grows wider as its advancing convection cell grows in intensity, height, and breadth.

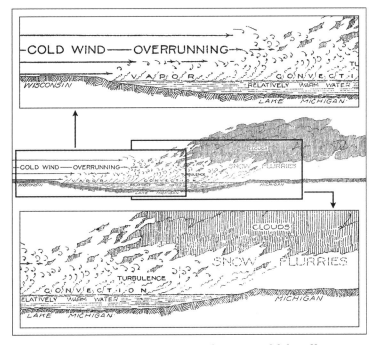

3.1. Wilfred Day's 1921 graphic explanation of lake-effect snow, enhanced with two overlapping detail images because the original drawing (7.5 inches wide and 2.7 inches tall) would not fit legibly on the page.

Forecasters, of course, need to know when to expect the polar air that drives the snowband. Because high pressure pushes air toward a region of low pressure, a weather map's isobars, which describe pressure trends, can explain—and in a sense predict—wind direction. In 1928, in another two-page *Monthly Weather Review* article, Robert M. Dole (1884–1966), an assistant meteorologist at the Lansing, Michigan, office, used a hypotheti-cal weather map to demonstrate the general relationship between pres-sure, wind, and lake-effect snow.[6] His map of "typical pressure conditions favorable for snow squalls" (fig. 3.2) reflects a low-pressure cell centered over Ontario, off the map to the northeast. Tiny arrows portray a counter-clockwise flow of air, a response to the Coriolis force, which deflects wind to the right (in the Northern Hemisphere) as it moves inward toward the center of low pressure. In late fall or winter, nearly north-south isobars like those shown here will direct cold northwest winds across a substantially warmer lake surface, producing "enormous clouds . . . very similar to thun-derstorms in appearance." But instead of heavy rain, downdrafts of cold air reach the surface "in the form of snow squalls." Dole didn't mention fetch, but it's clear that these northwest winds are favorably aligned to pick up lots of moisture from the lake.

Understanding a configuration is easier than predicting its occur-rence. Even so, an experienced forecaster who understood the lake-effect recipe would be on the lookout for an emerging low-pressure system likely to produce northwest winds. Dole was an old-style forecaster who studied sequences of past weather maps in hope of matching the current situation with historic storm tracks and previous patterns of highs and lows, which would then play out in a nearly identical fashion. The Weather Bureau had yet to accept the newfangled notion of a polar front and air masses, introduced in Norway around 1920, and predicting weather was more art than science.[7]

What a difference a decade can make: by the late 1930s Norwegian theory was well established in American meteorology, thanks in part to weather balloons with radio transmitters, which gave forecasters a more immediate picture of temperature and moisture at different altitudes. For Laurence W. Sheridan (1907–1984), radio soundings were the key to a pio-neering analysis of the vertical structure of a undistinguished snowstorm

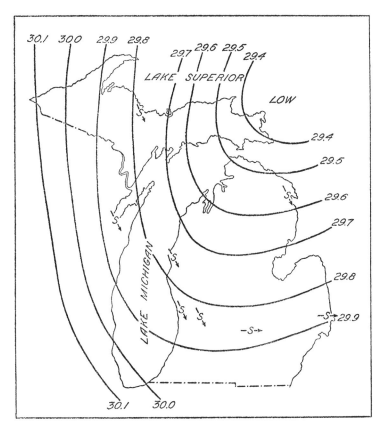

3.2. In 1928 Robert Dole used this map of pressure and winds to describe typical conditions for lake-effect snow in Michigan's Lower Peninsula.

chosen because better data were available than for other, more severe lake-effect events. Born in Buffalo, Sheridan received a PhD in mechanical engineering from Catholic University in 1932 and weathered the Great Depression with a string of teaching jobs, including a professorship in mathematics at the College of Mount St. Vincent, in New York City, a convenient locale for studying meteorology at New York University, which awarded him a master's degree in 1941. His three-page article "The Influence of Lake Erie on Local Snows in Western New York," in the December 1941 issue of the *Bulletin of the American Meteorological Society*, is probably based on his master's thesis—a footnote tagged to the title identified it as an "abstract of a longer paper," but I never found a fuller article.[8] Sheridan never published anything else on lake-effect snow, and NYU's archives do not include master's theses. Even so, his short article has the focus, substance, and significance a graduate advisor likes to see in a master's paper.

Sheridan's study focused on a low-pressure system that developed over the Great Lakes on March 16, 1941, and lasted four days. Centered over the St. Lawrence Valley, the system produced steady westerly winds across Lake Erie as confirmed by isobars and winds similar to those on Dole's map, but farther east. Radio soundings taken at Sault Ste. Marie on March 18 and at Buffalo the following day revealed the lake's effect on air currents circulating counterclockwise around the low and moving across the largely unfrozen lake. At an altitude of 0.2 kilometers the air was 9°C warmer after crossing the lake, and the specific humidity had increased by nearly 200 percent. But these differences fell off with increasing altitude, and at 4.4 kilometers there was no appreciable difference in either temperature or moisture. What's more, the air's "equivalent potential temperature," a measure of relative stability, showed a consistent increase with altitude at Sault Ste. Marie but an ominous decrease between 0.4 and 1.6 kilometers over Buffalo, indicating that the lower part of the air mass had become "convectively unstable" after crossing the lake and was thus vulnerable to further convective lifting when forced upward by hills rising 1,000 feet or more about the lake. It was no surprise that south Buffalo received 5 inches of snow, and higher areas to the southeast even more.

In addition to noting that wind direction "has a marked effect on the location of the heavy snow area," Sheridan observed that surface winds stronger than 25 miles per hour not only increased evaporation but created a "frictional turbulence" that further warms the base of the air mass and distributes moisture "through a fairly thick layer of the atmosphere." His paper concluded with perhaps the earliest recipe for lake-effect snow, complete with numbers but customized for an area he knew well.

> The factors, then, that combine to produce the type of storm here
> described are: a well-developed low pressure system which remains

stationary for several days, over the St. Lawrence Valley or north of the Georgian Bay region, causing cold air to stream southward and recurve across the Great Lakes bringing westerly winds to western New York; a wind velocity of at least 25 miles per hour; a temperature difference of close to 20°F between the air and lake temperatures; and, finally, a wind direction that varies little with altitude.

Although seldom cited, Sheridan's short article probably led to his next job, as a research mathematician and long-range forecaster at Weather Bureau headquarters in Washington, where he worked from 1942 to 1946, before moving back to academia as a mathematics teacher.[9]

By odd coincidence, the next issue of the AMS *Bulletin* carried another short article on Lake Erie snow by another western New York native, John T. Remick (1918–1961), who had written a bachelor of science thesis in meteorology at the Massachusetts Institute of Technology in 1941 and was identified as a 2nd lieutenant in the army air corps from Lockport, New York (on Lake Ontario, north of Buffalo).[10] Remick had enlisted as an aviation cadet three days after Christmas 1940, a little less than a year before Pearl Harbor.[11] Posted to MIT, which trained numerous military meteorologists before and during World War II, he rose to the rank of lieutenant colonel in the air force, and joined the Weather Bureau in 1946. At the time of his death in 1961, he was a meteorologist in the Emergency Warning Section, in Washington, DC. The John T. Remick Observatory, established by his father in his honor at Lockport High School, occasionally hosts meetings of the Buffalo Astronomical Association. Like Sheridan, Remick never published anything else on lake-effect snow.

Unlike Sheridan, who had focused on a particular air mass related to a single run-of-the-mill, late-winter snowstorm, Remick probed a decade of data in search of factors underlying the region's more extreme snowfalls. His analysis identified three distinct influences: frictional, thermal, and orographic. Friction is important because water is less resistant to airflow than land: moist air moving off the lake slows down, piles up, and cools as it's forced upward, often resulting in condensation and precipitation. While friction "alone tends to produce light continuous snow," the passage of cold air over relatively warm water and the resulting instability "tends to produce showers of snow in light to heavy amounts." Remick reasoned that the thermal influence "is at a maximum in the fall when the air is much colder than the water" and noted that not one of the severe Lake Erie storms he examined had occurred after December 16. The orographic influence is important because "the forced ascent of the air up a slope may be sufficient to start off precipitation [where] none would occur otherwise."

Remick's analysis caught the attention of Victor Paul Starr (1909–1976), who cited it as "a more detailed study of the Lake effect" in his 1942 textbook *Basic Principles of Weather Forecasting*.[12] That Starr mentioned Remick, not Sheridan, might reflect the former's focus on more extreme events, of obvious interest to forecasters. Or it might involve a bit of academic logrolling insofar as Starr had received his master's at MIT in 1938 and surely knew Remick's advisor, Hurd Willett (1903–1992).[13] Starr had joined the faculty at the University of Chicago's newly established Institute of Meteorology, and Chicago, MIT, Cal Tech, and NYU—where Sheridan had studied—were the only graduate departments of meteorology in the country. Motives aside, Starr's sense of lake snow seems more fully developed than Remick's, and in keeping with the chapter title "Forecasting the Actual Weather," he identified the critical role of wind direction in determining where squalls and "violent convection" were likely within "a narrow belt next to the shore."[14] "The success of forecasting for shore points," he wrote, "depends to a great extent upon the success with which surface wind direction can be predicted"—Dole's point entirely.

Predicting wind direction was especially frustrating at the Weather Bureau's Chicago office, where forecasters had to depend on sea-level prognostic charts, or "progs," prepared in Washington by the National Weather Analysis Center (NAWAC). Progs were an atmospheric crystal ball of sort: a map of isobars, adjusted for elevation differences and intended to show the pattern of atmospheric pressure thirty hours ahead. According to Chicago forecaster Lawrence Hughes (1918–2008), the progs were often wrong, typically portraying a clockwise flow that turned out to be counterclockwise when actual surface weather was mapped a day or two later.[15] In a letter published in December 1957 in the "Forecasters' Forum" section of the monthly house newsletter, *Weather Bureau Topics*,

Hughes argued that "since NAWAC is concerned with a fairly large area in prognosis, their knowledge and application of the climatology of each forecast district is not likely to be as complete as [that at individual] stations within each district." Simply put, as guidance for local forecasters concerned with Lake Michigan's effects on winter weather, Washington's maps were untrustworthy.

Hughes was not the first to imply that the forecasters' crystal ball had hit a glass ceiling. In the massive, 1,343-page *Compendium of Meteorology* published by the AMS in 1951, Hurd Willett, who wrote the lead chapter in the weather forecasting section, complained that "in spite of all this great expansion of forecasting activity, there has been little or no real progress made during the past forty years in the verification skill of the original basic type of regional forecast, of rain or shine and of warmer or colder on the morrow, the kind of forecasting which first received attention."[16] Though he had titled his chapter "The Forecast Problem," the solutions offered—more analysis, increased standardization, better training—seem obvious yet hardly trivial. Willett's pessimism was confirmed by data on the average accuracy of the 30-hour surface prog, which did not rise above 40 percent until 1964.[17]

In the early 1950s, at least a few research meteorologists were confident that numerical models of atmospheric circulation would soon offer forecasters more reliable guidance. While Willett conceded that "high-speed computing devices [placed] the feasibility of lengthy numerical reckoning on an entirely new basis," he was not confident of an imminent breakthrough. "In spite of initial optimism," he warned, "it is generally recognized at present by those who have been working on these methods that no radical advance in practical weather forecasting in the near future is probable."[18]

Jule Charney (1917–1981), who contributed the *Compendium*'s chapter on computer-based modeling, was more buoyant.[19] He began by praising the insights of Norwegian meteorologist Vilhelm Bjerknes (1862–1951), who had recognized as early as 1904 that mathematical equations could deal with the angular momentum produced by the earth's rotation while describing the horizontal and vertical motions with which the atmosphere transferred heat energy from warmer to cooler regions, and British physicist Lewis Fry Richardson (1881–1953), who "proposed to integrate the equations of motion numerically and showed exactly how this might be done." Because "the science of meteorology has progressed to the point where one feels that at least the main factors governing the large-scale atmospheric motion are known," Charney argued, "it seems clear that the models embodying in mathematical form the collective experience and the positive skill of the forecaster cannot fail utterly."

Richardson had outlined a strategy in his 1922 book *Weather Prediction by Numerical Process*: use the mapmaker's grid of meridians and parallels to partition a continent or hemisphere into discrete rectangles; extend these rectangles skyward as three-dimensional prisms; divide the prisms vertically at altitudes of 2.0, 4.2, 7.2, and 11.8 kilometers into stacks of three-dimensional cells; initialize the model by carefully estimating temperature, pressure, humidity, and winds at the center of each cell; and use the various equations of Newtonian physics, hydrodynamics, and thermodynamics to calculate the transfer of air and energy between adjoining cells.[20] The simulation was to move forward in time in discrete steps, and because heat energy, air molecules, and momentum could not mysteriously appear or disappear, each step required multiple rounds of readjustment, to let each cell respond to transfers to or from its horizontal and vertical neighbors. Though intriguingly clever, Richardson's model was so computationally challenging that his only trial run was based on a much abbreviated grid for which he calculated pressure and wind velocity at alternating cells (fig. 3.3). Despite specialized coding forms designed to expedite the process, calculations for a single six-hour interval took six weeks and produced absurd results.

Charney attributed Richardson's failure to poor data and the impracticality of hand calculation. With denser observations of surface weather and upper-air conditions well established in the postwar era and "the development of large-capacity high-speed computing machines" well underway, meteorologists were once again interested in Richardson's solution.[21] Although his *Compendium* chapter is short on details, Charney and some of his colleagues at the Institute for Advanced Study in Princeton, New Jersey, had been running test predictions using the ENIAC (Electronic Numerical Integrator and Computer) at the army's Aberdeen Proving

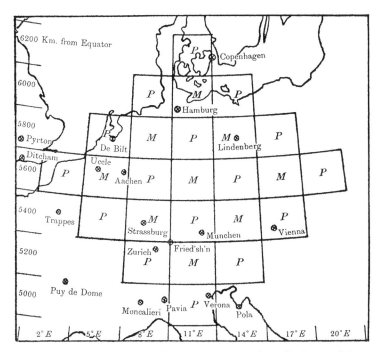

3.3. The simplified grid used by Lewis Richardson for his trial fore-cast consisted of only 25 cells. He further simplified the model by calculating only two parameters: momentum (wind velocity) for cells labeled M and pressure for cells labeled P.

Grounds in Maryland. Although the results were encouraging, a twenty-four-hour forecast based on a two-hour time step took twenty-four hours of computing time. The apparent bottleneck was the ENIAC's limited internal memory, but the Institute would soon acquire a new computer able to store 1,024 forty-bit numbers—a puny 5K of memory by today's standards—that promised to reduce the time to a mere half hour. "It is thus not entirely quixotic," he mused, "to contemplate the preparation of numerical forecasts for practical use in the near future."[22]

Charney had been at the Institute since 1948, when mathematician and computing pioneer John von Neumann (1903–1957) hired him to direct the Meteorology Project, part of a government-funded attack on computationally intensive problems of interest to the military. Predicting

the future of a vastly complex system like the atmosphere required what
Norman Phillips (1923–), a colleague of Charney's during the early
1950s, called "intelligent simplification" because it not only reduced
computational complexity but helped the scientist better understand
the problem.[23] Although Charney had worked out the mathematics for
a three-dimensional solution, he coped with the ENIAC's limitations by
programming a two-dimensional, single-layer model—computationally
akin to treating the atmosphere as flat.[24] In 1951 Phillips introduced a two-
layer model that afforded a crude but useful representation of the vertical dimension.[25] Two years later Charney and Phillips used the Institute's
new computer to produce "2½-dimensional forecasts" using an enhanced
single-layer model deemed "noticeably superior" to Charney's earlier version.[26] As computers became more powerful and more widely available in
the 1960s, 1970s, and 1980s, three-, six-, and twelve-layer models accorded
the atmosphere's vertical dimension an ever more accurate treatment.

Faster computers and more frequent, denser measurements of upper-
air conditions also contributed to the improvements in forecast accuracy
touted by the National Centers for Environmental Prediction (NCEP),
successor to the central forecast office at the National Weather Service, as
the Weather Bureau was renamed in 1970. Figure 3.4, based on an "operational forecast skill" graph from the NCEP, shows a generally upward
trend since 1955 for the thirty-six-hour 500 mb map of North America—
one of the local forecaster's most useful tools—and since 1976 for the
more ambitious seventy-two-hour forecast. Computer-based forecasting
has been operational (rather than merely experimental) since 1955, and a
skill level of 70 (not really a percentage) indicates a highly useful prediction.[27] I've added labels showing related advances in modeling technology
and data gathering. Ensemble forecasting, the most recent enhancement,
assesses the reliability of a prediction by running different models or the
same model using slightly different initial conditions.[28] Because a model's
performance depends heavily on how well the computer knows the state
of the atmosphere at the beginning of the simulation, an ensemble is useful in assessing a prediction's sensitivity to small amounts of error.

Numerical modeling was a fully operational guidance tool by 1979,
when Lawrence Hughes, now chief of the Central Region Science Services

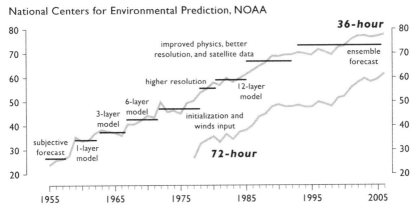

3.4. Improvements in numerical weather prediction reflect markedly faster computers as well as advances in modeling and observation technologies. Vertical axis represents "operational forecast skill," as measured for the 500 mb North American forecast map by the National Centers for Environmental Prediction.

Division, in Kansas City, published a study of forecaster prowess at sixty-six National Weather Service (NWS) offices in his region.[29] Hughes focused on precipitation, which is less easily forecast than temperature, and looked at thirteen years of verification data. In NWS parlance, *verification* is a management tool for assessing job performance, promoting employees with the best records, and devising training programs to improve overall performance.[30] It is based on skill scores that not only measure how frequently an individual forecaster gets it right but also compare his predictions to average weather for the location and time of year—an employee whose forecasts are no better than "climatology" might well be replaced by a set of statistical tables.[31]

Hughes concluded that routine verification helped forecasters sharpen their skill by providing useful feedback as well as the motivation "to obtain a better understanding of meteorological processes, . . . to make better use of observations, especially of radar and satellite, and to better understand the strengths and weaknesses of . . . numerical and statistical guidance."[32] He also cited lake-effect snow as the likely explanation for noticeably depressed average skill scores at stations where "snow flurries" are frequent, especially in the fall. Lake-effect snow was problematic because of the difficulty of distinguishing between small but measurable amounts of precipitation and merely "trace" amounts. "There is little a forecaster can do to improve this situation," Hughes noted, "although studies to separate trace events from measurable events *might* help."[33]

I have yet to find a study of the trace-measurable problem, but a search of the American Meteorological Society's electronic bibliographic database suggests that lake-effect snow gained a tenuous foothold in the meteorological literature by the 1960s followed by an uptick in the mid-1970s and a renaissance of sort from about 1979 onward, thanks to radar and satellite imaging, which helped forecasters better visualize the phenomenon. That's the trend apparent in figure 3.5, which juxtaposes three histograms showing the number of journal articles with "lake-effect," hyphenated or not, in the title, the abstract, or anywhere in the text.[34] A similar search using "snow flurries" found a few additional pre-1960 articles but none more perceptive than those already discussed.

The earliest article in the graph was disappointing—for both me and its authors.[35] In 1959 the eminent Norwegian meteorologist Sverre Petterssen (1898–1974) and University of Chicago researcher Philip Calabrese published a "preliminary survey," funded by the air force, of the effect of the Great Lakes on winter patterns of pressure and precipitation. Although their *Journal of Meteorology* article no doubt heightened awareness of lake-effect snow within the research community, it did little more than enrich the discussion with technical jargon like *lapse rate, nonadiabatic cooling, pressure field,* and "maximum relative vorticity" and give subsequent investigators something to cite. Its principal finding—that "detailed conclusions do not seem justified" because of "the sparsity of suitable observations"—can be viewed kindly as a call to look carefully at, and learn from, individual snowstorms.

Seven years later, *lake effect*'s debut in an article's title confirmed its place in the meteorological lexicon. "Mesoscale Study of a Lake Effect Snow Storm," published in the August 1966 *Monthly Weather Review,* focused on a two-day, early February snowstorm over Lake Ontario and demonstrated how a network of fourteen anemometers, nine instruments

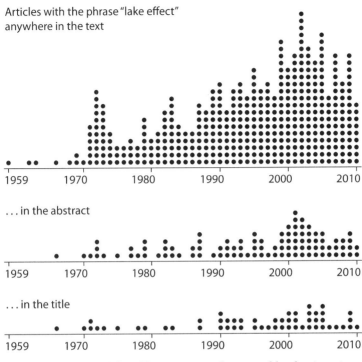

Articles with the phrase "lake effect"
anytwhere in the text

1959   1970   1980   1990   2000   2010

...in the abstract

1959   1970   1980   1990   2000   2010

...in the title

1959   1970   1980   1990   2000   2010

3.5. Occurrences of "lake-effect" in journals covered by the American Meteorological Society's online bibliography. The search included the *Monthly Weather Review* since 1871 but not the first fifty years (1920–1969) of the AMS *Bulletin*.

for recording temperature and humidity simultaneously, and forty-nine recording barometers could provide a detailed picture of surface conditions.[36] Co-authored by Robert Peace, a research meteorologist at the Cornell Aeronautical Laboratory, in Buffalo, and Robert Sykes, a college professor in Upstate New York, the article also demonstrated the value of collaboration, including in this case the observation network of the Atmospheric Sciences Research Center at the Albany campus of the State University of New York (SUNY), radar images from the National Weather Service in Buffalo, and wind and temperature measurements from a regional electric-power company. Peace and Sykes described the coordinated movement of parallel snowbands, which raised more questions than their data could answer. They attributed heavy snowfall rates to convergence—that is, the movement of surface winds inward, toward the snowband's axis—which made the area contributing heat and moisture "far larger than the area of the storm or individual air trajectories would imply."

Few people have studied lake-effect snow as intensively as Bob Sykes (1917–1999), who oversaw weather stations in Greenland during World War II. He retired from the air force in 1961 and taught meteorology until 1983 at SUNY's College at Oswego, where he focused on atmospheric processes affecting local weather, in particular the heavy snowstorms that immobilized Oswego in 1958, 1966, and 1972. Sykes supplemented whatever upper-air data and historical reports he could find with detailed on-site observations of sky conditions, snow depth, and timing—former students reported that he "actually entered [data] into his log books at 15-second intervals during heavy snowstorms."[37] He was especially intrigued by the *snowburst*, a local term used in the literature since at least 1950 to describe snow falling at two or more inches per hour from a lake band that held its position for several hours.[38] He would have applauded the 2010 National Science Foundation grant that gave SUNY Oswego faculty and students use of a Doppler-on-Wheels mobile radar station, similar to those used to study tornadoes.[39]

Sykes was a regular presenter at the Eastern Snow Conference, an annual meeting of meteorologists and emergency managers. The 1972 conference, held in Oswego, was memorable because a third of the hundred attendees who had gathered on February 3 for the two-day meeting could not leave until February 6 because roads out of town were blocked by several feet of new snow from what Sykes termed a "blizzardburst"—unlike a snowburst, in which lateral winds are minimal, this lake-enhanced storm was accompanied by strong winds and severe drifting.[40] Well before the conference returned to Oswego in 1992, organizers had prudently moved the meeting to late spring.

Another New Yorker who contributed to a fuller understanding of lake-effect snow is James Jiusto (1929–1983), a faculty member at SUNY Albany. In a 1972 article based on snow-depth and water-content data collected over a three-year period by seventy-five cooperative weather observers in New York, Ohio, and Pennsylvania, Jiusto and a co-author found that wind direction and timing accounted for important differences in storm intensity and "snowmass."[41] In addition to confirming that November and early December snowstorms produced generally heavier accumulations than January snowstorms, they discerned three categories of intensity, represented by the trio of exemplars mapped in figure 3.6. Intense snowstorms were comparatively concentrated, with up to 20 inches of snow in places, whereas moderate storms distributed lesser amounts of snow over a wider area. Both categories typically reflected a southwest to west airflow 5,000 feet above the surface, with the more intense storms having a steadier storm axis, a single snowband, and stronger convergence. By contrast, weak snowstorms, covering a wider area with accumulations no greater than 15 inches, typically reflected northwesterly winds, a more limited trajectory over a warm lake, and multiple snowbands. In a true lake-effect snowstorm, they concluded, wind direction and the difference in temperature between air and water were more important than orographic influences.

Jiusto's interest in lake-effect snow was probably piqued by conversations with Ronald Lavoie, a fellow PhD student at Penn State in the mid-1960s. While Jiusto's dissertation focused on cloud physics, Lavoie simulated lake-effect storms with a three-layer computer model.[42] His grid of 2,000 points around Lake Erie was much finer than those used to predict synoptic-scale storms, typically larger than 200 kilometers (120 miles). To better represent lake-effect snowstorms, which were smaller, more intricate "mesoscale" systems, Lavoie aligned the grid with the long axis of the lake and spaced its points 12 kilometers apart in the long-axis direction and 6 kilometers apart in the perpendicular direction (fig. 3.7, *upper*). To consider orographic influences, his data included a generalized map of local terrain.

Lavoie was generally pleased with the results for a trial simulation of the well-documented thirty-hour snowstorm of December 1–2, 1966, when cold arctic air behind a cold front crossed the unfrozen lake and snow fell at 1 to 2 inches per hour in places. Juxtaposition of his predicted and observed precipitation maps (fig. 3.7, *center* and *lower*) shows that the model more or less accurately predicted both the amount and the location of heavy snow but underestimated the extent of lighter snow southeast of

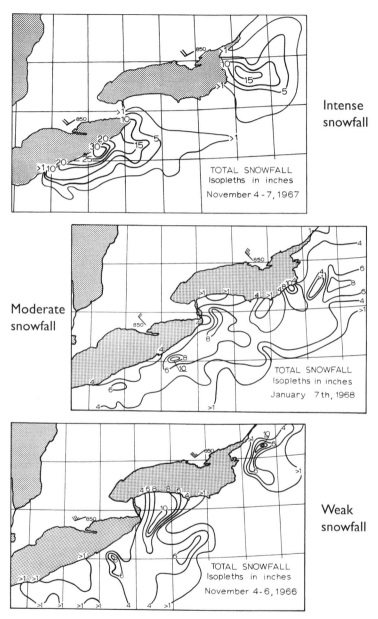

Intense
snowfall

Moderate
snowfall

Weak
snowfall

3.6. Maps comparing the snowfall and extent of typical intense (*upper*), moderate (*center*), and weak (*lower*) lake-effect storms.

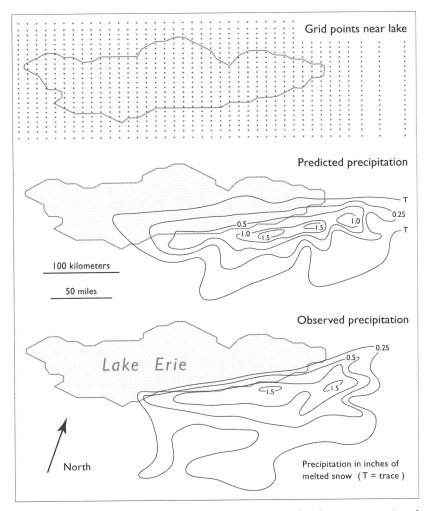

3.7. Maps of the portion of Ronald Lavoie's grid around Lake Erie (*upper*) and the predicted (*center*) and observed (*lower*) patterns of precipitation for the December 1–2, 1966, snowstorm, shown here as its water equivalent, in inches of melted snow.

the lake. Although the simulation was "intended more for diagnosis than for prediction," Lavoie noted, it "provide[d] a useful first-order approximation" of "surface forcing factors," and demonstrated convincingly that the difference in temperature between the lake surface and air aloft was "the dominant influence."[43]

Numerical weather prediction had yet to offer anything close to this level of geographic detail in 1987, when Tom Niziol, in an article in the newly established journal *Weather and Forecasting*, described operational forecasting at the NWS office in Buffalo. The National Meteorological Center, outside Washington, was using two models to create guidance products for local forecasters, but the newer one had a mere 15 to 20 grid points over the Great Lakes, and its older cousin "only one to two."[44] A few of these guidance products were useful parts of the Buffalo office's broad mix of forecasting tools, which included the "decision tree" mentioned in chapter 1. Weather radar fostered short-term predictions of the movement of lake-effect storms, over the next few hours, but could not monitor snowbands more than about 90 kilometers (56 miles) beyond the tower. Satellite imagery could fill in some of the gaps in radar coverage, but Niziol eagerly awaited mesoscale models with the higher resolution needed to predict the local impact of lake-effect storms.

Model resolution was only slightly better by 1995, when Niziol co-authored a follow-up assessment of lake-snow forecasting with colleagues at the NWS offices in Albany and Binghamton. A finer grid, with 38 points over the Great Lakes, was still "insufficient to explicitly forecast the formation and evolution of individual snow bands."[45] Even so, the authors were confident that numerical modeling might eventually predict storms able to dump a foot or more of snow in one area but less than an inch five miles away. Because models with vastly improved resolution would not work well unless initialized with vastly improved data, they called for "a high-resolution (temporal and spatial) observational network" that included automatic weather stations able to report surface temperature, pressure, and precipitation in real time; "wind profilers" able to estimate wind speed and direction at the various levels; and a new Doppler radar system that could locate snowbands 100 kilometers (60 miles) away. Older radar instruments could detect snowbands at almost the same range but revealed little about the intensity of snowfall or atmospheric motions within regions of precipitation.

The National Weather Service had a lot riding on its WSR-88D radar—WSR stands for "weather surveillance radar," 88 refers to the system's launch in 1988, and the D (for Doppler) indicates an ability to estimate wind speed and direction as well as the amount of precipitable moisture.[46] Nicknamed NEXRAD, for Next Generation Radar, the new system was the linchpin of a broad restructuring, or "modernization," of weather service operations. Pitched to Congress as a strategy for more timely and reliable severe weather warnings, the plan reconfigured forecasting operations around radar sites spaced more or less evenly across the country, except in very sparsely populated regions or where rugged terrain hindered radar surveillance. Instead of a network of 50 statewide forecast offices and 230 smaller offices with no forecast responsibility, the modernized weather service would consist of 116 Weather Forecast Offices (WFOs) with sufficient staff to take full advantage of the new data-collection and display systems and also keep up with an ever-changing array of numerical models and related guidance products.[47]

Modernization made each WFO responsible for a County Warning and Forecast Area (CWFA) and gave forecasters greater autonomy.[48] Employees are encouraged to develop expertise in one or more specialties, such as climate, web page development, or winter weather.[49] Local management includes a meteorologist-in-charge (MIC); a Science and Operations Officer (SOO), who focuses on training, introducing new technology, and promoting research; and a Warning Coordination Meteorologist (WCM), to liaise with media, local emergency managers, weather spotters, and the public.

The map of WFOs and CWFAs around the Great Lakes (fig. 3.8) reflects the general west-to-east movement of weather in the middle latitudes. The Chicago, Detroit, and Milwaukee WFOs were relocated to the far western suburbs of their namesake cities, and the forecast areas of the Buffalo and Cleveland offices stretch eastward. In New York as elsewhere, the reorganization was dramatic. Rochester and Syracuse lost their small local offices, which heretofore had adapted forecasts from Buffalo to the local situation, and the Binghamton forecast office picked up responsibilities for Syracuse.[50]

Radar's ability to detect or monitor a weather system depends upon the storm's distance from the radar instrument as well as its "depth," that is, the distance between the top of the storm and the land below. Radar beams are aimed at an upward angle, away from the instrument, and

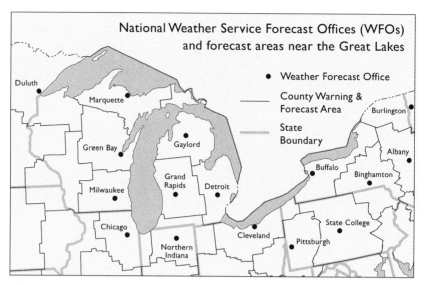

National Weather Service Forecast Offices (WFOs) and forecast areas near the Great Lakes

- Weather Forecast Office
  County Warning & Forecast Area
  State Boundary

Duluth
Marquette
Green Bay
Gaylord
Grand Rapids
Detroit
Milwaukee
Chicago
Northern Indiana
Cleveland
Pittsburgh
Buffalo
Binghamton
State College
Albany
Burlington

3.8. National Weather Service forecast offices and their areas of responsibility in the vicinity of the Great Lakes following modernization in the mid-1990s.

because the earth's surface is curved, at some distance the beam will overshoot a lake-effect snowstorm, typically shallower than two kilometers, and often lower than one. Because of public concern about possible degradation in radar coverage, the National Research Council (NRC) produced maps comparing NEXRAD with its predecessors for a variety of severe storms, including lake-effect snow.[51] For each type of storm the study panel estimated maximum ranges for the old and the new radars. Hurricanes could be spotted farther away than lake-effect snow, which in turn had a greater detection range than mini-supercell thunderstorms, and for all weather types NEXRAD had a longer reach than older radars. Figure 3.9, which compares coverage within areas vulnerable to lake-effect snow, indicated improved storm detection over most of the region but significant degradation in northeastern Indiana and northwestern Pennsylvania.

A companion map (fig. 3.10) showing old and new radar sites explains these changes. For example, improved coverage in the eastern part of the lake-effect zone reflects a new NWS radar site at State College,

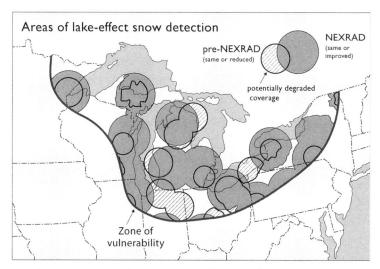

3.9. Coverage map in the 1995 National Research Council study comparing NEXRAD and pre-NEXRAD detection of lake-effect snow.

Pennsylvania (the + in the center of the state), and a new Department of Defense site at Griffiss Air Force Base, in Rome, New York (the × east of Lake Ontario), one of twenty-one military radars that supplement the NWS network. After Griffiss AFB was closed in the mid-1990s, the radar was relocated to Montague, New York, near the center of Tug Hill, where it covers the army base at Fort Drum, to its north, and is managed by the WFO in Burlington, Vermont.[52]

Like a brave soldier frustrated by red tape, the Montague radar is an underperformer. At 1,700 feet (0.52 km) above the lake surface, it's in what a 2007 report by a team of United States and Canadian meteorologists called "a prime location for detecting convective activity over and around the lake."[53] But because federal regulations require the beam to look upward at an angle not less than 0.5°, it often overshoots approaching lake-effect weather. A simulation suggests that reducing the lowest elevation angle to -0.4°—in effect letting the radar look downward—could "more than double" its range for detecting lake-effect snowstorms, to 135 miles (220 km), and "more than triple" its effective range for estimating snowfall accumulation, to 100 miles (160 km). Reducing the lowest elevation angle

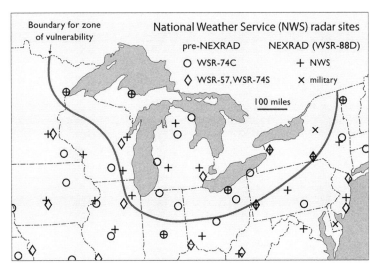

3.10. Map in the 1995 National Research Council study showing areas of potential degradation of radar-detection of lake-effect snow and the locations of old and new radars.

would require a costly environmental impact study, which might not over-come groundless public fear of radiation.[54] The small amount of radiation emitted by weather radar is orders of magnitude below levels that might affect human health and far less hazardous than a mobile telephone or microwave oven.[55]

The NRC maps (figs. 3.9 and 3.10) were bad news for northeast Indiana and northwest Pennsylvania, where modernization meant potentially degraded coverage after older radars were removed. Northern Indiana eventually received its own WFO, complete with NEXRAD radar—its omission could have been a glitch in the NRC study—but Northwestern Pennsylvania was not so lucky. The congressman whose district included most of the nine-county area served by the Weather Service Office in Erie, Pennsylvania, asked the US General Accounting Office (GAO, now the Government Accountability Office) to investigate. In September 1997 the GAO released a report that included a map (fig. 3.11), based on the 1995 NRC study, highlighting the resulting "coverage gap" for lake-effect

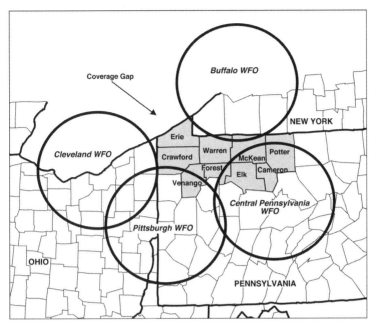

3.11. Nine shaded counties highlight the likely gap in radar coverage around Erie, Pennsylvania, because of plans to "spin down" the Erie Weather Service Office.

snow. Citing an internal NWS report, the GAO noted that "for about 35 percent of lake-effect snow events, the composite weather system will be insufficient to compensate for the degradation in radar coverage."[56] In reply, the weather service demonstrated no degradation for the nine counties overall and even suggested an additional radar might be installed. Despite the flap, the Erie office was soon phased out, or in NWS lingo, "spun down."[57]

In early 1999 the private sector filled the gap with the Hamot Doppler 24 Radar, a joint venture of television station WJET-TV, the local ABC affiliate, and the Hamot Health Foundation, which runs Erie's nationally prominent Hamot Medical Center. Hamot put up most of the nearly one million dollars for the project and received naming rights as well as improved weather reports for LifeStar, a medical transport service run jointly with another local hospital, and WJET upstaged WSEE-TV, the local CBS station, which had acquired a less sophisticated radar in 1988.[58] Erie's new radar was similar in principle to NEXRAD but far less powerful. Even so, it included software adept at the visual analysis of lake-effect snow, and because the data were processed on site, the imagery was more current as well as more locally detailed than NEXRAD imagery from Cleveland. It was just right for dramatic "nowcasting" with locally detailed maps showing areas at that moment, or soon to be, under a snow band. Viewers could also access the latest Doppler image of the region twenty-four hours a day through the station's website.

Radar is great for nowcasting, which focuses on readily observable features and looks forward no more than an hour or so, but ill-suited to forecasting, which relies on numerical models to look ahead at least thirty-six hours. Forecasting lake-effect snow depends heavily on predicting the temperature difference between the lake surface and upper levels as well as the direction of airflow, which can be frustratingly fickle or reliably stable for five days or longer. In making an informed guess, forecasters like Tom Atkins, chief meteorologist at WJET, have a range of guidance products at their disposal. When I asked Tom to identify the single most useful guidance product, he responded quickly with a single word, BUFKIT, the acronym for Buffalo toolkit.[59]

An interactive software package developed at the WFO in Buffalo in 1995, BUFKIT helps forecasters explore a variety of models and data sets, with a focus on atmospheric instability at a particular locality.[60] Its screen can display two or more graphs at once, or a single graph with a large menu. Half of the screen might contain a vertical profile, with temperature, moisture, or some other meteorological parameter scaled along the horizontal (X) axis and atmospheric pressure (a surrogate for altitude) along the vertical (Y) axis, while the other half is a menu for selecting models, parameters, and a time interval. The forecaster can use the cursor to select multiple profiles for display or point to a particular profile, which is then identified in the menu.

BUFKIT is not only a powerful visualization tool but also a gateway to frequently updated operational forecast models, including ensemble forecasts with up to sixteen elements and research models with grid points no more than 5 kilometers (3.1 miles) apart.[61] Originally devised to forecast lake-effect snowstorms, it was enhanced in 1997 to encompass a broad variety of meteorological events, and with support from the NWS Warning Decision Training Branch, in Norman, Oklahoma, it is regularly upgraded and used for operational forecasting, research, or training at more than seventy-five WFOs, universities, TV stations, and emergency management offices in the United States and Canada. And it's free.

However potent their technology, Great Lakes forecasters are understandably leery of exceptional snowstorms and heavy accumulations. Particularly thorny are out-of-season events like the "five-flake-plus" storm that dumped two feet of snow on the Buffalo area on October 12–13, 2006, when trees still had all their leaves. Dubbed Aphid—the theme for lake-effect snowstorms that season was insects—it began in the morning as rain but changed unexpectedly in midafternoon into heavy, wet snow, which brought down tree limbs and power lines, and continued past midnight with thunder and lightning.[62] Although Buffalo forecasters had accurately predicted the invasion of cold arctic air with west-southwest winds across a very warm Lake Erie, they considered conditions too warm for prolonged snow. Recognizing that something was amiss, they announced the likelihood of one to six inches of snow at 2:36 p.m. and issued a belated warning for "widespread tree damage and power outages" at 6:48 p.m.[63] Paradoxically, BUFKIT got it right. In a recapitulation of the incident, mortified meteorologists from the Buffalo office conceded "a calendar that read early October . . . most likely tainted a more objective analysis of the available guidance."[64]

Another noteworthy "surprise" highlighted numerical prediction's tendency to underestimate snow accumulation. On November 26–27, 1996, snowbands aligned with the short-fetch direction across Lake Ontario unexpectedly blanketed parts of western New York near Rochester with a foot of snow.[65] A numerical model had correctly predicted wind direction, but because north-south snowbands generally produce little lake-effect snow here, forecasters called for only a "dusting to one inch." The model had grossly underestimated precipitation because the storm's overwater fetch included Georgian Bay as well as Lake Ontario.

This "premodification" of cold air when it crosses an upwind lake can enhance the intensity and duration of lake-effect snow off downwind lakes as well as affect the location and timing of snowstorms.[66] Evidence for multiple-lake interaction includes satellite imagery with visible lake-to-lake cloud bands, which are more common than once believed. The most common lake-to-lake cloud bands run from Lake Superior across Lake Michigan and from Lake Huron across Lake Erie. Three-lake and even four-lake cloud bands have been observed, but the two-lake moisture plume is more common. Research on multiple-lake interaction might eventually inform operational numerical models.

For a firsthand look at recent applied research on lake-effect snow I attended the 19th Annual Great Lakes Operational Meteorology Workshop, held at Cornell University during two and a half days in mid-March 2011.[67] Half the presentations concerned lake-effect snow; a fifth dealt with warm-season weather, including flooding; another fifth addressed management issues such as warning systems, training, and forecast verification; and the remainder focused on severe weather's impacts on aviation. Most of the lake-effect papers were focused on either the genesis, forecasting, and impacts of specific storms or the use of data for multiple events to explore relationships or test hypotheses. Several presenters proposed new or modified indexes or rules for forecasting the onset or amount of snowfall—discussion underscored the need for fine-resolution models, the need to look closely at different models before issuing a forecast, and the lingering difficulty of predicting exactly how much snow a particular storm might produce at a particular place. A few presenters were university researchers, but most of the speakers and attendees were NWS forecasters or SOOs eager to learn from each other's experiences.

Whatever enhancements arise from these endeavors will further enrich the National Digital Forecast Database (NDFD), a dynamic atlas of forecast maps updated every hour as needed and covering a variety of weather elements, including predominant weather and amount of snow. Introduced in 2003, the NDFD treats the contiguous United States—"the lower forty-eight" if you ignore Hawaii's more southern latitude—as an array of grid points 5 kilometers apart.[68] Each WFO prepares forecasts for points within its area of responsibility, and software generates interactive national and regional maps, available online, as well as text and "voiced" products like local forecasts for NOAA All-Hazards Radio. NDFD predictions reach seven days forward for some elements, like temperature and predominant weather, but forecasters are unable to predict snow amounts more than forty-eight hours out. According to Tom Niziol, because the physical processes that produce precipitation are "extremely complicated . . . the best numerical models show very little skill more than 48 hours into the future."[69]

Figure 3.12 illustrates gridded maps used by the Buffalo WFO to display predictions of predominant weather, in this case, snow and less intense "snow showers," a day or more ahead of Lake Effect Storm Alewife, the first significant lake-effect event of the 2010–11 season.[70] The four maps show a persistent band off Lake Erie, which deposited as much as 10 inches of snow on the hills south of Buffalo, and a narrower band off Lake Ontario, which dropped up to 13 inches of snow on parts of Tug Hill before turning 45° by midmorning and drifting southward in the afternoon to dust the Syracuse suburbs. With symbols largely confined to the Buffalo forecast area, the maps told a familiar tale of warm lakes, cold air, and shifting winds.

Although reluctant to forecast beyond their own turf, forecast offices provide ready links on their websites to regional and national forecast maps and also post customized versions of "experimental" products developed elsewhere but deemed locally relevant. An example is the Experimental Probabilistic Snowfall Forecast for the Upper Peninsula, provided by the Marquette, Michigan, forecast office and based on technology developed by NWS scientists in Pontiac, Michigan, and Tulsa, Oklahoma.[71] An online array of 60 maps (fig. 3.13) divides three days into six 12-hour forecast periods ending at 8 a.m. or 8 p.m. On 42 maps in the first seven rows color bands portray estimated probabilities for 12-hour snowfalls exceeding 1, 2, 4, 8, 12, 18, and 24 inches. Additional maps in the lower three rows provide best-guess estimates of the new snow likely to accumulate during each of the six 12-hour periods as well as plausible maximum and minimum amounts for each period. Viewers

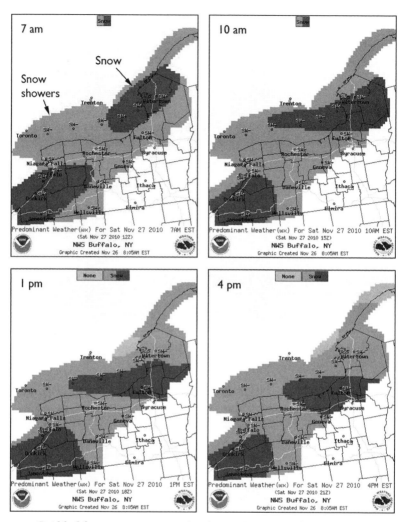

3.12. Gridded forecast maps posted online at 8 a.m., Friday, November 26, 2010, accurately described the predominant weather at 7 a.m., 10 a.m., 1 p.m., and 4 p.m. the following day.

can click on any of the 60 maps to examine a cartographic forecast like figure 3.14, on which color bands and percentage probabilities describe the chance of receiving at least an inch of new snow overnight.[72] An "82" on the map indicates that snowfall was most likely at Marquette, directly on Lake Superior.

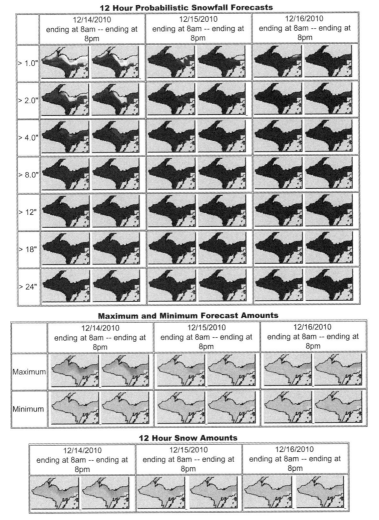

3.13. A 6 × 10 array of small maps show the likelihood of various amounts of snowfall over twelve-hour time periods as well as the plausible range of outcomes.

NDFD grids and interactive probabilistic forecasts are merely the latest in a series of innovations that pushed weather forecasting through two dramatic transitions: from largely manual methods to nearly full automation and from essentially verbal pronouncements to heavily cartographic

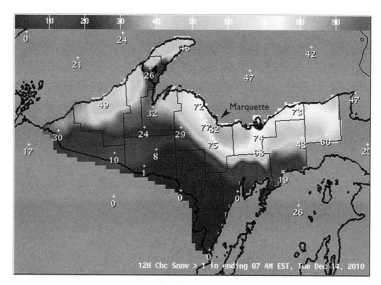

3.14. Forecast map obtained by clicking on the upper-left map on the web page featured in figure 3.13. Numbers indicate the probability of at least one inch of new snow during the twelve-hour period ending 7 a.m., December 14, 2010.

web-based presentations. Change has been both geographic and sociological, with Doppler radar precipitating the creation of WFOs and CWFAs, and the piecemeal development of NDFD grids requiring collegial consultation with neighboring WFOs as well as greater teamwork within individual offices.[73] Although electronics would seem to covet a role more central than mere guidance, humans will continue to run the system, rework the software, conduct research, and deal with the public.[74] The frontier of weather prediction is not only technological and institutional but also spatial and temporal, with improved resolution and a longer forecast horizon among its most needed enhancements. Simply put, a 5 kilometer grid is still too coarse for lake-effect snow.

4 Impacts

Few things grab a news executive's attention more than a "best" or "worst" list with the hometown at the top. When the list impugns local winters, a loyal editor naturally points a finger at other cities with colder temperatures or more snow. Not so, though, when the 2011 *Farmer's Almanac* (released in September 2010) ranked Syracuse number one on a list of five cities with the nation's worst winter weather. Mildly amused, the *Syracuse Post-Standard* labeled the dubious distinction an "accolade" and quoted a local booster who felt that "worst" really meant "best."[1] In 2002, the last time *Farmer's Almanac* (not to be confused with *The Old Farmer's Almanac*) tried to commandeer headlines by rating climates, Syracuse was smugly pleased to beat out Buffalo, scratched from the 2011 competition because its average snowfall was too low.[2]

Being the poster child for Upstate New York's snowbelts is fine as long as a city can cope, which Syracuse almost always can, along with Buffalo and the other three Great Lakes cities on the *Farmer's Almanac* list: Duluth (no. 2), Cleveland (no. 4), and Detroit (no. 5). Coping means knowing what to expect and having snowplows, salt spreaders, and trained operators ready when needed. Cities that anticipate lots of snow typically deal with severe winters far better than those unwilling to prepare for these inevitable but less common events.

Climate statistics that focus on averages can be dangerously deceptive, as demonstrated in February 2010, when back-to-back "snowmakers" paralyzed Washington, Baltimore, and

Philadelphia for more than a week.[3] Philadelphia endured a seasonal total (79 inches) nearly four times greater than normal (20 inches), and Baltimore was close behind, with 2009–10 and normal amounts of 77 and 18 inches, respectively.[4] (Brings to mind the fable about the statistician who drowned trying to wade across a river with an average depth of two feet.) Meanwhile, snowy Buffalo counted only 74 inches for the season, while Syracuse registered a below-average 106 inches.[5]

A storm's impact is not just a matter of how much snow falls and its moisture content: if strong winds cram snow into road cuts, a six-inch snowfall can shut down a major highway. When windblown snow reclaims a highway minutes after the snowplow has passed, it's a safe bet the storm is a blizzard, defined by the National Weather Service as a storm with "sustained winds of 35 miles per hours or greater, and considerable falling or blowing snow"—a dust storm with snow that can cut visibility to less than a quarter mile.[6] By this definition, the Great Lake snowbelts are hardly Ground Zero for North America's blizzards, which are most common in the Upper Midwest. Climatologists Robert Schwartz and Tom Schmidlin, who examined forty-one years of data, found a "blizzard zone" covering western Minnesota and the Dakotas, where counties averaged more than one extreme snowstorm a year.[7] About a hundred miles to the west is Casper, Wyoming, with fewer blizzards but just enough Rocky Mountain snow to capture the number three spot on the *Farmer's Almanac* worst winters list.

According to Schwartz and Schmidlin, blizzards affect twenty-six million Americans annually—almost a tenth of the population.[8] An average year sees only 10.7 blizzards, but the number has been rising. Fortunately for people in the northern half of the country, most snowstorms lack the sustained winds that characterize blizzards.

Although areas directly downwind from the Great Lakes fall outside the blizzard zone, snowstorms are more frequent here than elsewhere in the eastern United States. This pattern was confirmed by climatologists Stanley and David Changnon, father and son, who teamed with NOAA climate expert Tom Karl to analyze data collected at 1,222 weather stations between 1901 and 2001.[9] In focusing on snowfalls of a half-foot or more over one or two days—less dramatic than a blizzard but definitely disruptive in Baltimore and Philadelphia and no doubt noteworthy in Buffalo and Syracuse—they found substantial variation from year to year as well as from decade to decade, with an increased frequency during the 1910s and 1970s, and marked regional variation, mostly influenced by elevation. It is hardly surprising that their map of average annual number of snowstorms shows the highest frequencies in the West, where the Rockies, the Sierras, and the Cascades produce extreme orographic uplifting, while northern New England, the Adirondacks, and the Appalachians are prominent in the East. Equally apparent are isolines reflecting regionally high frequencies south of Lake Superior and east of Lakes Michigan and Ontario (fig. 4.1), which experience four or more snowstorms in an average year. By contrast, the average is 1.5 or less within the urban corridor between New York and Washington, DC, where winter often passes without any serious snow.

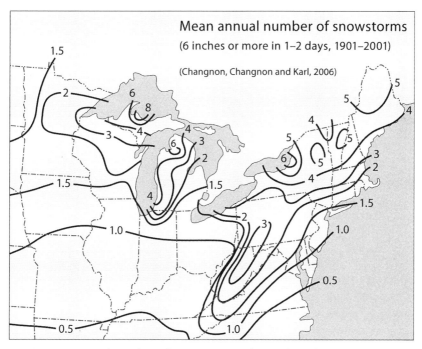

4.1. Average annual number of snowstorms with 6 inches of snow or more over one or two days, 1901–2001.

Noteworthy lake-effect snowstorms are more frequent than their coastal counterparts but affect smaller areas, often with a wide variation in accumulation over a distance of only 20 or 30 miles. Anchored to the source of moisture, bands of lake snow are usually narrow and shorter than 100 miles, in marked contrast to the huge synoptic-scale snowstorms that travel up the East Coast, not far from the ocean. To explore this hypothesis, I compared two sets of high-impact snowstorms, one based on the Lake Flake scale devised by the Weather Forecast Office (WFO) in Buffalo for the region east of Lake Erie and south of Lake Ontario, and the other defined by the Northeast Snow Impact Scale (NESIS) developed at NOAA's National Climatic Data Center (NCDC) for the Northeast urban corridor. Like the five-category Fujita and Saffir-Simpson scales used to rate tornadoes and hurricanes, both scales sort disruptive snowstorms into five categories.[10]

Let's start with NESIS, rolled out in 2004 but applied retrospectively to earlier high-impact snowstorms with suitably detailed data.[11] A storm's relative impact is first described by its NESIS score, which considers both snow accumulation and the population affected.[12] NESIS scores range from 1 to more than 10 for the most extreme storms, which paralyze transportation across a large, densely settled region. Based on its impact score, a storm is assigned to one of five categories named to denote increasing degrees of disruption: Notable, Significant, Major, Crippling, and Extreme. Only two of the forty high-impact storms listed on the NCDC website for the period 1956–2010 were Category 5 (Extreme) events.[13] The March 12–14, 1993, storm (fig. 4.2), which dumped more than 30 inches of snow on parts of six states, and over 10 inches on parts of fourteen more, registered the highest NESIS score, 13.20. By contrast, the pair of storms that shut down much of the East Coast in early February 2010 had scores of 4.38 and 4.10, which placed them two steps down, in Category 3 (Major).[14] Even so, two additional high-impact snowstorms—a Category 3 (NESIS score = 5.46) in late February and a Category 2 (NESIS score = 3.99) two months earlier, in mid-December 2009—helped break seasonal snowfall records in Washington, Baltimore, Philadelphia, and several other East Coast cities.[15]

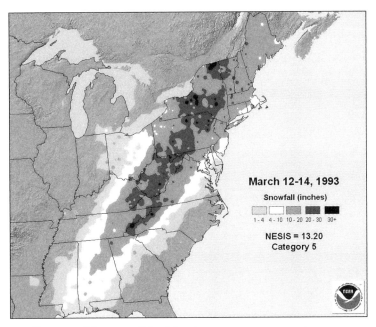

4.2. Extent and impact of the March 12–14, 1993, snowstorm.

Paradoxically, winter 2009–10 was comparatively gentle east and south of Lakes Erie and Ontario, where Buffalo forecasters named only six lake-effect snowstorms, the most significant of which was the five-flake Amaryllis storm—the theme that season was flowers—a mid-December lake storm described as a "classic" event produced by a west-to-southwest flow across both lakes, with "all ingredients prime" and snowfall extending "well inland."[16] Snowbands developed first off Lake Erie, in the wake of a large storm that had moved eastward from California, and bands formed off Lake Ontario several hours later. Amaryllis persisted for approximately 54 hours, and generally steady snowbands produced peak amounts of 27 inches at Springville, east of Lake Erie, and 40 inches at Highmarket, east of Lake Ontario on Tug Hill (fig. 4.3). Although the graytones in figure 4.3 cannot do justice to the Buffalo forecast office's impressive color map, the nested bands reflect heavier snowfall near the center and confirm the persistence of two very intense snowbands. Strong winds and whiteout

conditions closed the New York Thruway (Interstate 90) east of the Pennsylvania line, and stranded over a hundred motorists. The impact was
greatest in "the Southtowns" 10 to 25 miles south and southwest of Buffalo, where residents substituted vigorous snow shoveling for a workout at
the gym and praised the unknown inventor of four-wheel-drive vehicles.[17]
Tug Hill, with fewer residents, had even deeper drifts. Even so, the disruption was short-lived in both areas, as snowplows cleared roads, tow trucks
pulled cars out of snow banks, schools reopened, and lift operators wel-
comed avid skiers.

Like NESIS, the Lake Flake scale considers both population and
snowfall. In addition to naming storms and assigning ratings, the Buffalo WFO prepares both capsule and detailed summaries of each storm's
genesis as well as peak and representative snowfall totals and their effects

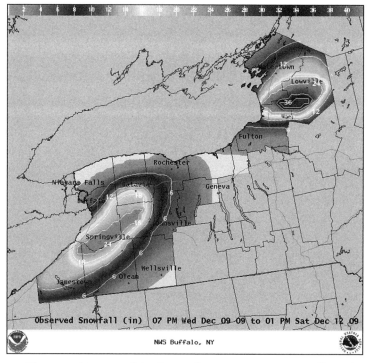

4.3. Extent and impact of the December 10–12, 2009, Amaryllis Lake
Effect Storm. Legend across the top reflects colors used to represent
different depths of snowfall.

on transportation and business. Between fall 1998 and spring 2010, Buffalo forecasters named a total of 120 lake-effect snowstorms, only seven of which merited five flakes. Over the same period, the NCDC reported twelve snowstorms with a NESIS score greater than 1.0, the threshold for Category 1 (Notable). There were no Category 5 (Extreme) storms during this period, and only two of the twelve qualified for Category 4 (Crippling) and only five for Category 3 (Major). What's more, only three of the twelve high-impact NESIS events overlapped any of the 120 named lake-effect snowstorms, and their respective maps show almost no geographic overlap: the NESIS storms left little, if any, snow near the lower Great Lakes, and each of the three lake-effect snowstorms lagged a day or two behind its NESIS counterpart.[18]

Seasonal influences explain the minimal overlap. November and December, when the lakes are relatively warm, have been peak months for lake-effect snow off Lake Erie, while January and February are the high season for the broad synoptic snowstorms that shut down East Coast cities and occasionally sideswipe Upstate New York.[19] Although lake storms occur from October into April, they have smaller footprints, attached to one or more of the lakes, whereas synoptic snowstorms in the East typically affect a much larger area as they move northeastward along the coast.

Anyone leery of generalizations will enjoy an hour with the maps and summaries on the Buffalo WFO's Lake Effect web page. As its capsule comments attest, few lake-effect storms are perfect textbook examples: only 8 of the 120 named storms were labeled "classic," while 12 were either "hybrid" or partly "synoptic," and another 10 were described as "lake-enhanced" or linked to a front. Quirks are rampant, as exemplified by phrases like "thunder reported," "heavy metro impact," "impressive multi-phased event," and "crippling out of season event." Although a warm lake and cold air are the basic ingredients, the recipe for lake-effect snow tolerates a delightfully diverse menu.

Delight hardly describes the feelings of a highway superintendent whose job depends upon a prompt, decisive response to snow-covered roads. Present-day hand wringing over snow removal makes it easy to forget that snow was less a problem than an asset until the early 1920s, when the gasoline engine banished the horse from everyday transport.

The coating of snow that means lost traction for wheeled vehicles once greased the flow of commuters along city streets, and in rural areas, where muddy or rutted roads were the norm, hard-packed snow made getting around much easier and helped lumberjacks move logs out the forest. As geographer William Meyer pointed out in *Americans and Their Weather,* continuous snow cover expedited heavy hauling in colonial New England and Upstate New York, while frequent thaws thwarted inland commerce farther south or closer to the coast.[20] Even so, urbanites came to see snow largely as refuse. As historian Bernard Mergen reminded readers of *Snow in America,* "City streets filled with horse manure and garbage were seldom fit for sleighing or walking."[21]

Neither wagons nor sleighs were ideal for long-distance inland transport. Turnpikes were costly and poorly integrated, and canals were slow, out of service during winter, and impracticable for many routes. Railroads emerged in the 1830s as a promising all-season alternative. From a handful of short, unconnected lines, the American rail network grew ever denser and reached its peak during World War I. Efficient operation demanded comparatively gentle grades, usually provided by a moderately devious route with alternating cuts and fills. Winds swept the fills and trestles clear of snow but easily plugged the cuts, making the line impassible and occasionally stranding a train, along with its crew and passengers. In addition to snow fences, designed to trap drifting snow before it engulfed the tracks, defenses included mounting a plow blade on the engine (for lesser drifts), running a special train with a snowplow car ahead of the engine (for deeper drifts), and sending a rescue train with a hastily recruited crew of trackmen and jobless laborers (for mega-drifts deep enough to ensnare a snowplow).[22] Though trains were unavoidably delayed, railroads managed to cope with blizzards, synoptic storms, and lake-effect snow.

Cast-iron wheels on metal rails also expedited local travel within cities, large and small. In the 1830s the horsecar began to replace the animal-powered omnibus, and by 1881, just before the electric motor afforded greater efficiency, 415 street railways with 3,000 miles of track and 35,000 employees were carrying over 1.2 billion passengers a year in 18,000 cars pulled by more than 100,000 horses and mules—which ate up much of the profit by consuming 11 million bushels of grain.[23]

Electrification offered new solutions to the perennial problem of snow removal, and when heavy snow taxed the competence of motorized plows and rotary sweepers, the traction companies hired platoons of day laborers with strong backs. Although streetcar snowplows that pushed snow off the tracks onto side roads and sidewalks angered teamsters and business owners, some municipalities saw the nuisance as an opportunity and made the trolley companies responsible for removing snow from streets with car tracks.[24] The deal fell apart when the rubber tire displaced the steel wheel. As historian Blake McKelvey noted in *Snow in the Cities*, Rochester taxpayers had to pay more for snow removal when the local transit company started converting car lines to trolley buses in the 1930s.[25]

McKelvey's book led me to an obscure 1917 report by the Rochester Bureau of Municipal Research, a think tank established two years earlier by philanthropist George Eastman, inventor of roll film and the Kodak camera.[26] The city's Department of Public Works, which was responsible for collecting garbage, cleaning streets, and removing snow, had asked the bureau to critique its snow-removal policies, with careful consideration of geography and the competing needs of motor vehicles and sleighs. Trucks had largely taken over the delivery of coal, and most of Monroe County's 15,464 automobiles and trucks were owned by city residents, who used them throughout the year. The growing number of motor vehicles had increased congestion, particularly in winter, when sleigh owners asserted a traditional right to use city streets. The report's authors agreed that "sleigh traffic from the country" had "no urgent business on Main Street" and proposed "to divert sleighs to certain streets [but] keep snow on certain crossings, to enable sleighs to cross without becoming stranded on car tracks." A map of downtown on which blue and red lines (gray and black in fig. 4.4) distinguished sleigh routes from "snow panning streets" describes their remedy. *Panning*, a careful reading reveals, meant pushing the snow through manholes into storm sewers, which led to Lake Ontario via the Genesee River.

Rochester is distinctive among Upstate New York's larger cities for plowing its residential sidewalks. The city took on this responsibility in 1905 because school children and other walkers were often inconvenienced by property owners who ignored a local law requiring prompt

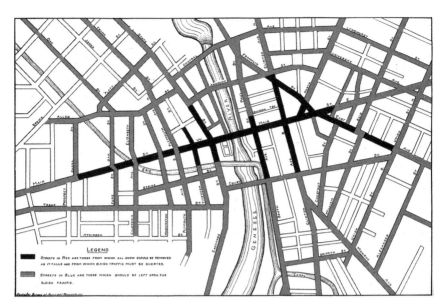

4.4. The Rochester Bureau of Municipal Research proposed keeping sleighs off the downtown portion of Main Street (*black*) and limiting "sleigh crossings" along car tracks.

shoveling after a heavy storm.[27] One or two scofflaws on a block could undermine the efforts of their more civic-minded neighbors, but enforcement was troublesome, and vigorous shoveling could kill elderly residents eager to avoid a fine. To keep the sidewalks open, the city hired contractors with horse-drawn plows, paid them by the linear foot, and added the cost to the property owner's tax bill. Two cents a foot per plowing was too generous, according to the Bureau of Municipal Research, which proposed an hourly rate for each plow team.[28] According to its 1917 report, an average season required thirty-five plowings—homeowners were still responsible for cleaning up after the light dustings common in lake-snow cities. It's unclear who cleaned up after the horses.

Although internal-combustion engines now provide the horsepower, the "embellishment fee" is still based on a property's front footage. According to the city's website, sidewalk plowing is merely a "supplementary service" to help property owners meet their "legal obligation" to clear adjacent sidewalks of "snow, ice, and snowplow residue."[29] Plowing begins after three inches of new snow has fallen, and the program covers 878 miles of sidewalk with a minimum width of five feet. An average "plow run" of nine miles takes five hours.

Rochester's commitment to pedestrians attracted a few converts, mostly small municipalities. For several decades Hamilton, New York, an affluent village forty miles southeast of Syracuse and home to Colgate University, has been clearing residential sidewalks with small tractor-plows that work outward from the local public school.[30] Sixty miles southwest of Buffalo, the small city of Jamestown used wooden plows on its hilly walkways until 1967, when six gasoline-powered sidewalk plows replaced twenty men and thirty horses.[31] In Buffalo and Syracuse, which have repeatedly rejected a similar service as too costly, unshoveled sidewalks, particularly after a heavy snow, often force pedestrians and people in powered wheelchairs into the streets.[32]

Buried fire hydrants become a hazard when street plows push snow off the road indiscriminately. Fire protection is a key justification for street plowing, but valuable time can be lost shoveling out a hydrant. Firefighters in the snowbelts do some of the shoveling themselves but also depend upon "adopt a hydrant" campaigns and the enlightened self-interest of nearby residents. Exceptionally heavy snowstorms might require help from the National Guard or a local prison. Rescuers know where to dig when the hydrant supports a spring-loaded vertical rod, two to four feet long, divided by reflective paint into red, white, and blue sections readily visible above the snow if you know where to look. Syracuse flags its hydrants by driving a metal post into the ground a foot or so away and topping it with a small red-on-white sign representing a fireplug (fig. 4.5). You know you're in serious snow country when the markers are more than six feet tall.

No less troublesome are parked cars.[33] Narrow streets become even narrower because snowplows cannot easily clear the pavement all the way to the curb and motorists avoid snowbanks by parking farther outward. Curb-to-curb plowing focuses on arterials essential for fire engines and ambulances, and even cities south of the Mason-Dixon line plan ahead by marking "Snow Emergency Routes," along which parking is banned whenever significant snow is expected. Arterial bans work best when motorists get the message before their cars are snowed in and the city has

4.5. Fire-hydrant marker in Syracuse, New York. The 6 × 7 inch sign is 6½ feet off the ground.

tow trucks poised to remove stragglers. Because built-up areas with few driveways or garages are vulnerable to clogged streets, Baltimore encourages residents to park free in municipal garages or on school parking lots.[34] Farther north where snows are more frequent, Syracuse relies on a combination of emergency snow routes, strategic plowing, a zoning ordinance requiring off-street parking for rental units, and a year-round odd/even parking rule that compels residents to move their cars to the other side of the street every day at 6 p.m.

In rural and suburban towns where streets are narrow and lack curbs, the only efficient way to clear snow before the morning commute is a total ban on overnight on-street parking.[35] In the towns surrounding Syracuse the ban runs from November through March or mid-April.

Getting onto the road is the homeowner's responsibility. Many residents, like myself, gladly pay a private contractor several hundred dollars a year to have our driveways plowed—an important off-season activity for local landscape service firms, which also plow parking lots at shopping centers.

Motorists nowadays seem less affected by heavy snow than half a century ago, according to weather historian Dave Call, who looked at four upstate cities (Buffalo, Rochester, Syracuse, and Albany) from the late nineteenth century onward.[36] A former weathercaster who teaches geography and meteorology at Ball State University, Call noted that "massive return-to-work traffic jams" were common in all four cities in the 1950s and 1960s. Uncertain why the day-after delays disappeared, he posited "a combination of population decline, job losses, and improvements in parking regulation" as well as more reliable, better quality equipment.[37]

Efficient snow removal demands not only heavy trucks with powerful engines and high-traction tires but also hydraulic plow blades that can be quickly turned or lifted. A highway plow with an adjustable wing blade extending up to ten feet outward to the right can clear a wide swath of roadway and maneuver around parked vehicles. Some highway departments put a second worker in the cab to operate the wing blade. In lake-snow country, where frequent snowstorms justify the investment, enterprising owners of pickups and large SUVs mount a hydraulic plow blade on the front and recruit a string of customers.

Grooming narrow residential driveways is an elaborate dance that includes inching up to the garage door with the blade raised; dropping the blade and backing away to scrape the surface while pulling the snow out to the street; and then pushing it to the side, off the roadway. Plowing snow into the street and leaving it there is illegal, and moving it across the road onto a neighbor's property typically requires the owner's permission. A conscientious contractor will stop back later, to clear the end of any driveway clogged anew by a municipal street plow making its rounds. Getting a larger helping of snowplow residue than one's neighbors is a frequent complaint.

Removing deep or even moderate accumulations requires a variety of equipment, including huge front-end loaders for lifting snow into a dump truck for transport to a nearby river, field, or unused parking lot. Where to put the snow can be a serious problem in cities, where merely shoving it to the side—onto sidewalks or into homeowners' driveways and side streets—angers residents, even during an emergency. Plans must include lists of prearranged dumping areas, both public and private, as well as pre-approved contracts with local construction companies that can quickly deploy additional equipment and workers. In lake-effect areas in particular, highway departments can't rest after the storm has subsided and the pavement has been cleared; their plows must push snow banks back even farther, off the shoulders of rural and interstate roads, to make room for new snow, certain to come.

Roads and Streets, a monthly magazine for civil engineers and contractors, chronicled the efforts of states and counties to keep their roads clear. A 1928 survey of equipment and practices observed that after "a few years [in which] snow removal activity has advanced by leaps and bounds in the 36 states where [it] is at all a problem," emphasis had shifted from new to replacement purchases "more in line . . . with corrected ideas [about] the most feasible methods."[38] Between the winters of 1922–23 and 1926–27, the miles of road from which snow was removed jumped from 27,096 to 106,721, while the average cost rose from $28.12 to $43.50 per mile. These figures, the editors asserted, reflected a "modern conception of snow removal [as] having at hand a plentiful supply of high-speed equipment, and operators subject to sudden call, day or night, as in a fire department." Photographs illustrated a variety of equipment in operation, all of it mechanized. Fifteen pages of ads for plows, four-wheel-drive trucks, snow loaders, and snow fencing interrupted the following year's update, which quoted a maintenance engineer for the Michigan highway department who argued, "the importance of open roads in isolated regions cannot be measured by the number of vehicles."[39] Because state officials considered commercial deliveries, physicians' house calls, and school buses absolutely essential, even roads with less than a hundred vehicles per day were plowed.

In Houghton County, Michigan, on the Upper Peninsula, drifts as high as fifteen feet hamper snow plowing. As late as winter 1938–39, a third of the 900 miles of county roads was kept open only for sleigh traffic.[40] By 1950 an expanded school-bus network required plowing 850 miles of road, including 112 miles of state highway under contract.[41] In an article for *Roads and Streets*, the county highway engineer noted two pressing needs: the replacement of obsolete snow-removal equipment and increased cooperation with neighboring jurisdictions. Maintaining his aging fleet of plows and graders was a costly distraction because of wartime shortages and austerity budgets, and many commuters and delivery vans, unlike school buses, did not stop at the county line.

Outlying towns and cities had become increasingly reliant on trunk highways since the early 1920s, when railways began discontinuing passenger trains and abandoning branch lines.[42] Although trains provided more reliable intercity connections during winter, cars and buses were undermining the profitability of rail service, which was withdrawn or reduced, putting greater pressure on highway departments to keep roads open.

Through traffic was a routine headache in Chautauqua County, New York, astride the Lake Erie snowbelt southwest of Buffalo. In a 1932 *Roads and Streets* article, county highway superintendent Squire Fitch (that's his real name) grumbled about motorists traveling US 20 without tire chains. They'd get stuck, abandon their cars, and seek refuge at the nearest farmhouse. "Clearing the road of cars takes longer than the plowing," Fitch complained, "and when the road is cleared, the cars have to be put back on the road."[43]

Demand for plowing varies widely in extreme western New York. As Fitch noted in a 1938 article in *Public Works*, "It is quite common for a blizzard to howl in the snow belt while golfers are enjoying rather mild weather in the southern part of the county."[44] Rather than assign each plow to its own district—standard procedure in many jurisdictions—a dispatcher kept track of plows and snow conditions with an unlisted phone number and a map on which white thumbtacks with numbers represented individual trucks and red thumbtacks indicated areas needing urgent attention. When a driver phoned in, the dispatcher issued a new assignment and moved the truck's thumbtack accordingly. Fitch, who had rejected radio dispatching as not only too costly but unreliable over distances greater than five miles, would surely be dazzled by GPS navigation and tracking systems, geospatial databases describing the precise locations of land boundaries and guard rails, and software for reconfiguring plowing routes in real time.[45]

Fitch apparently had little faith in road salt and salt spreaders, conspicuously absent from his detailed descriptions of plow blades and truck lights. Even so, his passing mention of "the purchase and hauling of cinders" indicates that Chautauqua County was spreading abrasives, at least sporadically, to increase traction on hills and other slippery spots. Fifteen years later and two hundred miles east, Syracuse was relying heavily on salt—but not sand or other abrasives—to cope with frequent lake-effect snowfalls of three inches or less. Writing in *Public Works*, city road czar Frank Harmon praised equipment manufacturers, chloride producers, and a dedicated workforce for helping him bring "Florida streets to Syracuse" in the winter.[46] He also noted that sand and cinders clog sewers and "make a messy looking street between storms" but conveniently ignored the accelerated body rust that after a few years readily distinguished a Syracuse car from its Florida counterpart. Although the nickname Salt City reflects the commercial salt industry that was prominent from the 1790s through the 1920s, many residents think it refers to aggressive snow removal.

States and localities can tell you what they spend on road salt, but the total cost of snow and ice control is often elusive. In DeWitt, New York, where I live, road salt, regular wages, and overtime accounted for 31, 49, and 17 percent, respectively, of the $938,122 spent in 2009 on "snow removal," which in turn represented 22 percent of the total cost of maintaining and improving the town's highways.[47] The true cost is a lot larger, though, because employee benefits, machinery, and fuel are hidden in other parts of the budget. My hunch is that snow removal probably costs town taxpayers more than $1.5 million annually, which works out to about $63 for each of DeWitt's 24,000 residents—quite reasonable, actually, considering the unmistakable impact and unavoidable necessity of winter plowing. To assure accountability, the town highway superintendent must stand for election every four years, along with the town supervisor and members of the town board, whose jobs are likewise in jeopardy should snow removal prove inadequate.

Efficient snow clearing demands cooperation among municipalities, counties, and the state, which are responsible for specific parts of the road network. In New York State, the Department of Transportation buys 950,000 tons of salt annually; uses more than 1,400 large dump trucks and a winter staff of 3,500 operators and 460 supervisors to plow 35,380 lane miles of roadway; and pays counties, cities, and towns to clear another 7,620 lane miles.[48] A 1946 amendment to the state highway law let the state outsource snowplowing to counties and municipalities, and counties sought similar savings by outsourcing some of their work to towns and villages.[49]

Onondaga County, which includes DeWitt and Syracuse, might someday subcontract all of its snow and ice control.[50] Whether this would save money is questionable insofar as subcontracted snow removal has a long history of higher-order governments subsidizing lower-order units.[51] A likely consequence would be an increased need for motorists to pay close attention to road signs marking municipal boundaries. I recall vividly my "ice accident," several decades ago, when the road curved abruptly as I crossed from Bethlehem, New York, which cleared its roads religiously, into Albany, which did not. Boundaries matter.

Timing also matters. Highway departments try to clear the roads before the morning commute and remove new snow before the evening rush hour. Persistent but generally light lake-effect snow accommodates this schedule, but intense lake bands that shift abruptly do not. Because highway officials need to know the likely intensity of a storm as well as its onset, they sometimes supplement National Weather Service forecasts with customized snowstorm predictions by private meteorologists who specialize in estimating duration, accumulation, moisture content, and temperature as well as changes that might require calling in additional workers or switching tactics.

Paradoxically, more detailed and reliable government forecasts have created a demand for still greater customization. In 1926 G. R. Thompson, who oversaw snow removal in Detroit, worried that rising temperatures might require a prompt clearing of gutters so that melt water from a wet snow would not pond in the streets.[52] "Contrary to popular belief," he maintained, "predictions of the Weather Bureau are almost always correct." Correct perhaps, but not as tailored to snow plowing as many highway superintendents would like. In the early 1950s, for instance, Frank Harmon, who promised Syracuse residents "Florida streets" in winter, hired a private forecaster to predict when (or whether) he'd need to stop salting and start plowing.[53] A 2010 National Research Council study identified local transportation departments in need of "very specific information to plan for clearing ice and snow" as a key "niche market" for private meteorologists.[54] The National Weather Service, their report concluded, has neither the budget nor the personnel to offer highly detailed, customized forecasts.[55]

The same study pointed out the difficulty of forecasting local differences in the amount of snowfall as well as the position and sharpness of the rain/snow line separating snow or sleet from rain or freezing rain.[56] Highway officials are especially leery of ground temperatures hovering near freezing, which raise the possibility of black ice, so-called because it is transparent as well as treacherous. To address these concerns, highway engineers have begun to embed sensors in the pavement and bridge decks and to integrate measurements of temperature, moisture level, and snow accumulation with government weather data.[57] However promising this new (and expensive) technology, rural highway departments in lake-effect country rely largely on NWS forecasts and reports from plow operators in the field.

Knowing when (or whether) roads will be cleared is especially important to superintendents of rural and suburban school districts, where almost all students are bused. State laws specify a minimum number of instruction days, usually 180, and most districts in lake-effect regions adapt by adding several snow days to their calendar—better to plan ahead than having to forego spring vacation or extend the school year into late June.[58] Fortunately, a school day delayed an hour or two still counts, as does a day shortened by an early dismissal. Highway departments and school administrators have learned to understand each other's needs and capabilities as well as the concerns of parents, prone to worry about late buses and lost wages.

Some school districts accommodate parents' desire for a midwinter Florida or Caribbean getaway by scheduling two one-week vacations, one in February and the other in April, rather than a single spring vacation in March, which is typical farther south. In addition to making snowbelt living more bearable, the winter break is one less week for superintendents to fret about snow closings.

Money spent on efficient snow control is money well spent, as demonstrated by the Lake Erie snowbelt's response to exceptionally heavy snowfall in December 1989, when districts between Cleveland and Buffalo received 50 to 80 inches of snow, twice the normal amount and 90 percent of it before the Christmas holiday. Geographer Tom Schmidlin, who surveyed thirty-nine school superintendents, deemed the impact "minimal," despite increased overtime and equipment rental costs related to snow removal from parking lots and access roads.[59] Eighteen districts reported closures, up from four the previous December, but sixteen shut down for only a single day, and the remaining two canceled only two days of classes. This successful response reflects careful planning as well as a bit of luck, for as one superintendent noted, much of the snow fell on weekends, which diminished the disruption.

In deciding whether to cancel or delay classes, school superintendents rely heavily on weather forecasts: if conditions are bad in the morning and likely to worsen later on, there's little point in bringing students in for a few hours and then sending them home. I canvassed the modest literature on school administration for guidance on snow-closing decisions but found little beyond the need for planning, reliable contacts, an understanding of community expectations, and a carefully worded letter telling parents where to find announcements.[60] In the magazine *School Administrator*, Jack Turcotte, a school superintendent in Maine, advised closing for the season's first big snowstorm, when "all the crazies overreact."[61] For guidance, he relied on superintendents in neighboring districts; contacts at the National Weather Service, the Federal Aviation Administration, and the state highway department; and a paid weather consultant. In the same issue Randy Dewar, an education professor who had worked for twenty years as a superintendent in the Midwest, warned against delaying the decision or delegating the responsibility to someone else.[62]

Closer to the Great Lakes, the city school district in Wooster, Ohio, south of Cleveland, outlines its snow-closing strategy on its website.[63] Officials are wary that a delayed start can be especially stressful for families and that reversing a decision because of worsening conditions can endanger students. When uncertain what to do, it's often best to cancel classes. State rules allow districts five "calamity days" a year for snow closings, flooding, power outages, and other emergencies.

Because parents need to know when classes are canceled or delayed, school superintendents rely on local radio and television stations to get the word out, and local media cooperate because timely school announcements help them build and retain an audience. And because well-structured lake-effect snowstorms are largely predictable over the next half-day, the geography of delays and closings is also an ad hoc weather report. Having learned the names and approximate locations of school districts in Central New York, I can listen to a radio report and construct a mental map of snow bands off Lake Ontario. But I wouldn't know the names of half the districts were I not interested in local weather.

The lake effect's impact on public education is typified by Lake Effect Storm Carp, a four-flake, five-day event that deposited up to four feet of snow on parts of Central New York during the first full week of December 2010.[64] On Monday morning, snowbands driven by northwest winds targeted areas west and north of Syracuse as well as uplands thirty miles to the east, where two rural districts canceled classes and a third declared a two-hour delay, to ensure that snowplows arrived before the school bus. My map (fig. 4.6) shows only two other closures, one for the Syracuse City district most students walk, sidewalk shoveling is spotty, and unplowed side streets are a hazard.[65] Suburban districts west and north of the city, where most students are bused, opted for a one-hour delay, just to be safe. The absence of closings along the Lake Ontario shoreline reflects a comparatively mild storm requiring an added boost from higher elevations.

Tuesday's closings map was less focused. A multi-band storm propelled by westerly winds interfered with school transportation along a broad arc running from the rural towns west of Syracuse to remote areas northeast of Tug Hill as well as along the southeastern shore of Lake Ontario. On-air announcements available around 7 a.m. (fig. 4.7) were

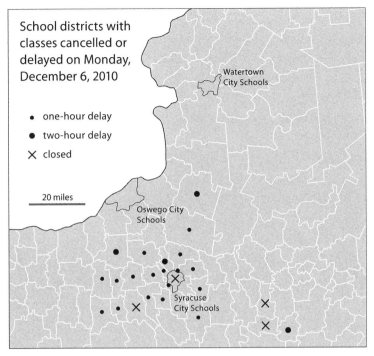

4.6. School closings and delays in Central New York on the first day of the early December snowstorm.

dominated by two-hour delays, with complete cancellations in a quartet of more elevated districts south of Syracuse and one-hour delays on the storm's western margin. In rural districts around Syracuse, the delayed opening is a common response because highway departments often need just a bit more time to catch up with the overnight accumulation. Not that day, though: by 9 a.m. superintendents within the snowbands had sensed the cumulative consequences of prolonged snowfall and canceled classes altogether (fig. 4.8). Wary that a delayed opening might escalate into a full closure, many parents (as well as pupils eager for the season's first "snow day") stayed tuned to their favorite local news station, which increased the drama by reporting traffic accidents and other impacts. Syracuse schools opened normally, but the city school districts in Oswego and Watertown, which serve substantial outlying populations, joined with their neighbors and closed altogether.

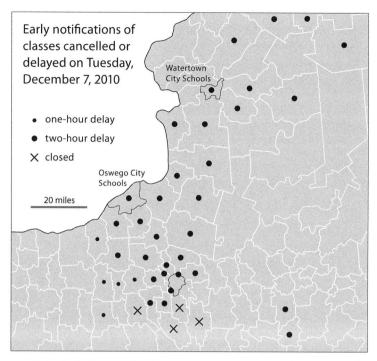

Early notifications of
classes cancelled or
delayed on Tuesday,
December 7, 2010

Watertown
City Schools

• one-hour delay
● two-hour delay
✕ closed

Oswego City
Schools

20 miles

4.7. School closings and delays as announced around 7 a.m. on the
second day of the early December snowstorm.

By Wednesday morning the Lake Ontario snow machine had
regrouped into a long, narrow snowband running through Syracuse and
affecting school districts twenty-five miles to the west and forty miles
to the east (fig. 4.9). Closure was the predominant response, with a few
one- or two-hour delays at the edges. This tight, cigar-shaped pattern of
cancellations is emblematic of lake-effect snowstorms, which are smaller,
more focused, and often more locally enduring than their synoptic counterparts. By Thursday, though, the lake effect had shrunk to light dustings
and snow flurries.

For insights on school closings in Michigan's Upper Peninsula, I interviewed Hank Bothwell, the recently retired superintendent of the NICE
Community School District, which sprawls across 680 square miles west
of Marquette and serves 1,200 students and their families.[66] Unlike his
counterparts in Upstate New York, Bothwell never ordered a one- or

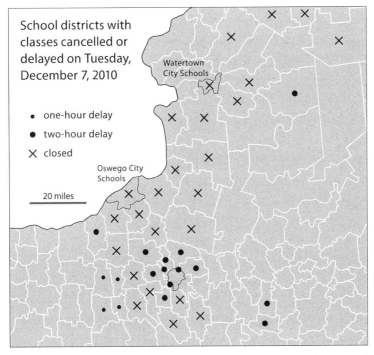

School districts with classes cancelled or delayed on Tuesday, December 7, 2010

Watertown City Schools

• one-hour delay
● two-hour delay
✕ closed

Oswego City Schools

20 miles

4.8. School closings and delays as announced by 9 a.m. on the second day of the early December snowstorm.

two-hour delay, which would cause serious problems for working parents as well as students in vocational and special-education programs shared with adjoining districts. "Big problems" would occur if the districts were "out of sync."

On days when significant snow seemed likely, Bothwell would get up at 3:30 a.m., check NOAA weather radio or the website of the WFO in Marquette for an update, and call the district's transportation director, whom he'd then join for a twenty-minute ride over a predetermined route, to "get the feel of road conditions." Home again, he'd talk with two neighboring superintendents as well as the dispatcher at the county highway garage before checking back with his transportation director about the parking lots at the district's two big buildings; normally plows would start clearing the lots at 3:30 a.m. The goal was to have the decision out on local radio by 5:30. Usually there was a consensus among those consulted. In

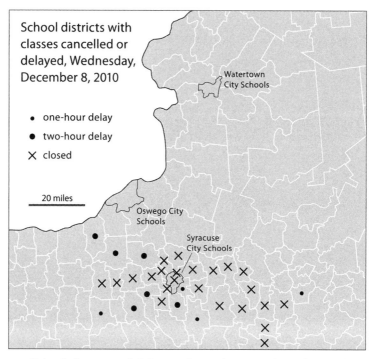

4.9. School closings and delays in Central New York on the third day of the early December snowstorm.

some instances, closing was a foregone conclusion: when NOAA issues a blizzard warning, "you'd be a fool to run school."

Because of state regulations and issues with teachers unions, closing school could be costly, especially before Michigan replaced its 180-day minimum with a 1,098-hour minimum. In most years a district can meet the new requirement by adding extra minutes here and there to make up for missed instruction days. Because the state requires a full accounting, the change means more paperwork for the superintendent's secretary.

Although canceling school is disruptive—most businesses must stay open even when schools close—parents' tolerance for having school when the weather's bad has declined in the past two decades. Bothwell blames TV weathercasters, "who hype it up and make storms look threatening." And pupils eager for a "snow day" no doubt amplify the threat.

For most families going anywhere is hardly an option during exceptional events like the late January 1966 storm that dumped eight and a half feet of snow on Oswego and closed 348 miles of the New York Thruway, where 1,500 stranded motorists sought shelter at a single service area.[67] A model of command and control, the Thruway was better prepared than most local governments. A private weather forecaster advised on when to shut down which section; a teletype connected toll booths, maintenance garages, and road-patrol stations to a "war room" at Thruway headquarters in Albany; the Red Cross had stockpiled cots and blankets at service areas; and helicopters helped evacuate the seriously ill.

Limited access at toll stations makes it easy to shut down parts of the Thruway where motorists might become stranded. The importance of selective closures was underscored by Lake Effect Storm Bluegill, a five-flake, three-day event that brought over three feet of snow to parts of Buffalo in December 2010.[68] On the evening of December 1, rain turned to snow and jackknifed trailers halted traffic along portions of the Thruway where tolls had been removed several years earlier to appease local commuters and elected officials. Drivers could readily enter the free zone at numerous on-ramps where toll collectors might otherwise have halted access and reported the gridlock. As snow off Lake Erie continued to fall, hundreds of motorists were trapped in their cars, some for more than seventeen hours. The ensuing finger pointing led to plans for local police, firefighters, and state troopers to block entrance ramps quickly.[69]

Dave Zaff, the Buffalo SOO, outlined these proposals the following March at the Great Lakes Operational Meteorology Workshop.[70] While shutting down an interstate highway seems preferable whenever an intense snowfall threatens a backup lasting two or more hours, closing a road takes time and diverts first responders from other duties. Each exit ramp will have an "owner" responsible for exiting traffic, and on-ramps will eventually have automatic snowgates like those at railway crossings— many interstates in the Midwest already have them.[71] If motorists begin to drive around the gates, plowing in the on-ramps is an option, but as Zaff warned, much of the traffic will merely clog other roads. And drivers of big trucks built like "motels on wheels" often "don't care if they get stuck."

Few localities in the Great Lakes snowbelts enjoyed a high level of emergency preparedness until the 1980s, when heightened concern over nuclear accidents and toxic chemicals forced states and counties to reconfigure their Cold War–era civil defense units as emergency management offices, in charge of coordinating response to a broad range of natural and technological disasters. How seriously they appreciate winter weather as a threat worth comprehensive planning depends on location and the current political climate. In a mid-1990s survey of New York's county-level emergency managers, severe winter storms were rated the second most serious hazard, just behind flooding, but in a geographically broader Federal Emergency Management Agency study, flooding and nuclear power were tied for first place while winter weather went unrated.[72] In 2001 the unexpected 9/11 attack on the World Trade Center precipitated a radical reorientation, and terrorist threats eclipsed natural disasters and technological hazards on emergency planning agendas throughout the state.

Despite continued antiterrorist efforts, New York's Office of Emergency Management still addresses a broad range of threats. Its current Multi-Hazard Mitigation Plan—a recipe for lessening impact before disaster strikes—includes a ranked list of six types of projects particularly relevant to snow and ice storms:[73]

1. Public awareness
2. Hazard-resistant construction
3. Tree pruning
4. Strengthen/improve/enforce building codes in hazard areas
5. Retrofit critical structures
6. Redundant utilities/communications

Little can be done to reduce the need for snow removal, but pruning trees and shoring up vulnerable structures can minimize the impact of an unprecedented winter storm. Perhaps the Great Lakes snowbelts' greatest defense against severe winter weather is a high level of public awareness reinforced by advertising initiatives as well as the repeated assaults of lake-effect snow. The news media never let us forget about our quirky snow and its possible impacts.

5 Records

Record snowfall and *snowfall record* aren't the same thing but they're close enough to give a chapter titled *Records* a coherent theme. *Record* is an obvious adjective for an exceptionally deep and disruptive snowfall, but without consistent and comparable *records*, its assumed preeminence deserves to be questioned. Because wind moves snow around, where you insert the measuring stick can make a difference, even within a short distance. A weather observer eager to talk of big snows experienced firsthand could easily spin the data by plunging the probe into the deepest spot. And even though a heavy snow might not be the storm of the century, it could qualify as the area's biggest November storm, the biggest in the past fifty years, or even the biggest in the new millennium.

Official snowfall data let cities face off in an annual competition that combines the come-from-behind excitement of horse racing with the right to brag about the region's biggest snowfall this season. In Upstate New York the annual snowfall race typically begins in early November, when the region's larger cities (Buffalo, Rochester, Syracuse, Binghamton, and Albany) get their first dustings of Great Lakes snow. Progress is erratic: one city pulls ahead only to be passed by another the following week as winds shift or Atlantic moisture intervenes. If the race is close, the winner might not be known until late May, when further measurable snow is highly unlikely. In addition to media acclaim, the victor gets temporary custody of a trophy called the Golden Snowball.

The upstate snowfall race started in the late 1970s, when Peter Chaston, a National Weather Service meteorologist in Rochester, rammed a foam ball onto an old Little League trophy, spray-painted it gold, and presented the award to the city that finished the season with the most snow—a good way, he thought, to "add a light touch" to a string of harsh winters.[1] In a typical winter Buffalo takes an early lead but falls behind Syracuse after Lake Erie freezes over. Even so, the fickleness of upstate winters let Buffalo, Rochester, and Binghamton each hold the trophy at least once between 1978 and 1984. The contest withered in the mid-1990s, after the NWS transferred Chaston to Kansas City and closed its Rochester and Syracuse offices.

In 2002 meteorologists in Buffalo revived the competition, and Syracuse-area Web developer Patrick DeCoursey offered an annual cash award of $100 to a children's charity in the winning city and rolled out a new trophy, crafted locally at the A-1 Trophy Shop (fig. 5.1).[2] Former Syracuse mayor Lee Alexander is alleged to have kept the original ornament, and a cleaner accidentally totaled its replacement, on permanent display at the Buffalo weather office.[3] An unabashed Syracuse boaster-booster, DeCoursey runs the New York State Golden Snowball Award blog, where he updates snowfall totals for the five cities and chronicles the impacts of new snowstorms. A table compares each city's running total for the current season with its normal-year and previous-season tallies as of the same date—like runners in a marathon, the cities compete against past performance as well as against each other.

GoldenSnowball.com also posts historic snowfall records, which extend the contest back to 1940–41, when Buffalo began a six-year winning streak.[4] Syracuse and Binghamton couldn't compete until 1951–52 and 1952–53, respectively, because comparable snowfall data were lacking. With Syracuse registering the most snow for forty-two of the fifty-eight years between 1952–53 and 2009–10, DeCoursey feels justified in disparaging Buffalo's seven victories during the contest's first twelve years by tabulating "undisputed" and "disputed" wins separately.

Other upstate cities would like to compete but are excluded because they're too small. In 2004, after Oswego's mayor pointed out that his city's snowfall consistently puts Syracuse to shame, DeCoursey initiated

5.1. The Golden Snowball trophy, photographed in December 2010 at Syracuse City Hall. Although the related website includes snowfall records for Albany and Binghamton, the only cities mentioned on the faceplate are Buffalo, Rochester, and Syracuse.

a Silver Snowball Award, with a $50 donation to a kids charity in the snowiest small city.[5] The first year's winner, Fulton, with 147.7 inches, bested both Syracuse (137.6 inches) and Oswego (131.4 inches) as well as DeCoursey's other two smaller cities, Utica (93 inches) and Watertown (70 inches). It's not clear why Binghamton, with only 47,380 residents in the 2000 Census, is a Golden Snowball contestant, while Utica, with a 2000 population of 60,651, is not. Even so, Binghamton not only hosts one of upstate's three Weather Forecasting Offices but had enough snow to beat out the other four Golden Snowball cities in 1963–64, 1966–67, and 1982–83.

No matter what the state or region, the most extreme snowfall records usually accrue to tiny, remote places sufficiently elevated to intercept huge amounts of lake-effect or synoptic snow: places like Montague, New York, fifty miles north-northeast of Syracuse, in the middle of Tug Hill. Montague attracted national attention when a local weather observer reported that 77 inches of snow had fallen between 1:30 p.m., January 11, 1997, and the same time the following day—an inch more than the previous twenty-four-hour record, measured on April 14–15, 1921, at Silver Lake, Colorado, seventeen miles due west of Boulder and two miles east of the Continental Divide. Weather junkies were excited that a lake-effect snowstorm in New York State had snagged a national record for extreme snowfall long associated with the Rocky Mountains.

Climatologists in Colorado were neither convinced nor amused. Skeptics included Nolan Doesken and Arthur Judson, who had extolled the Silver Lake record in their book on measuring snowfall.[6] Judson, a retired snow scientist with boyhood memories of Tug Hill snowstorms, conceded he'd be "delighted" if the Montague record were "the real thing."[7] But Doeskin, who was assistant state climatologist, told a Syracuse journalist he'd "be personally offended" if Silver Lake's record were broken.[8] He planned to request a review by the National Weather Service because 77 inches of snow in twenty-four hours was "on the limit of what nature can do."

There were good reasons to question the Montague measurement. After all, cold air holds far less moisture than warm air, and compaction would surely shrink a deep accumulation. In 1938, pioneer snowfall cartographer Charles Brooks made a rough calculation based on absolute humidity and snow density and concluded, as a "conservative" estimate, that "the maximum possible depth of snowfall in one day might be considered to be approximately six feet," five inches short of the 77-inch Montague record.[9] That his theoretical maximum depth was only an approximation was underscored by a 1953 *Monthly Weather Review* article in which a Weather Bureau scientist declared the 76-inch Silver Lake measurement "reasonable."[10] In noting that the 1921 Colorado snowfall was markedly less dense than Brooks's hypothetical accumulation, in which 10 inches of snow contained an inch of water, the article emphasized the point that light, fluffy snow can accumulate more rapidly and to a greater depth than denser snow—a point that made the Montague measurement at least plausible.

As official arbiter of US snowfall records, the National Weather Service responded to Doesken and other skeptics by promptly appointing a committee of six scientists, who reviewed relevant data and visited Montague in early February to inspect the site and interview the volunteer snow spotter. The committee, which included Doesken and the state climatologists for New Jersey and New York, paid particular attention to the observer's notes and measurements and the role of the Buffalo Weather Forecast Office in recruiting and training a dense network of snow spotters and in handling their reports. The committee's report, released in March, included carefully worded findings and recommendations that mixed praise for the network's worthy aims with criticism of lax practices and a call for corrective measures.[11]

The key finding was no doubt a disappointment for Tug Hill residents perversely proud of their new national record. Although the committee concluded that the Montague snowstorm "was indeed a very large snowstorm," its report recommended "the 77-inch total not be recognized as an official climatological snowfall amount for that 24-hour period."[12] Though this might seem like double-talk, the recommendation reflects an important distinction between the operational needs of forecasters, who rely on volunteer snow spotters to help them understand the often-quirky behavior of lake-effect snowstorms, and the more exacting scientific standards for climatological records.

What undermined the 77-inch record was its construction from six measurements for periods ranging from 1 to 12 hours. Five of the measurements were 3.5 hours or less, in clear violation of a rule not to measure "more than once every 6 to 12 hours."[13] Wary of melting, Buffalo meteorologists had also cautioned snow spotters, "Do not measure every hour and them add them up [because] this would give an unrealistically high amount." A somewhat different irregularity did not invalidate the 76-inch Silver Lake record, interpolated from a measurement of 87 inches, recorded after 27.5 hours, but the report questioned whether this estimate would be confirmed if measurement guidelines in use at the time were "rigorously applied."[14] The committee pondered devising a statistical process to better compare the Montague and Silver Lake figures but scrapped the idea because it "would add an additional layer of uncertainty to observations that already involved some estimation."[15] A single 24- or 27.5-hour measurement clearly trumped a summation for six shorter intervals.

The committee had no qualms about the observation site. A photo in the report (fig. 5.2) confirmed that the snowboard was in "an excellent location for snow measurement, open, not affected by buildings, and with enough nearby trees to result in little drifting, even during relatively strong winds."[16] The principal problem was that the observer, who was "extremely enthusiastic about taking snow measurements and reporting to the forecast office in a timely fashion," cleared the board after each measurement, thereby reducing the amount of compaction. Although the instruction sheet said nothing about whether or when to clear the board, snow spotters were encouraged to phone in reports anytime snow was falling at more than an inch an hour.[17] The committee lamented the lack of water-equivalent (melted snow) measurements as well as the careful documentation required at official climate stations, but pronounced the volunteer spotter "knowledgeable and competent."[18]

5.2. The Montague snow spotter's measurement site included a 2-foot-square snowboard and two 6-foot snow stakes.

Committee members sought additional insight by interviewing snow-plow crews, mapping snow totals reported by other spotters, and calling up archived satellite images. A veteran plow operator attested that the January 11–14 snowstorm, with snow falling at up to 5 to 6 inches an hour at times, was the largest he had seen in twenty-five years. Other operators concurred that the deepest accumulations had occurred near Montague but noted that the snow had settled quickly, as is typical of light, fluffy lake-effect snow. A four-day map of total new snow compiled by Buffalo meteorologists from reports by other observers revealed a peak accumulation around Montague and slightly to the south (fig. 5.3). A weather satellite image concurred by showing a marked lake-effect band over Tug Hill. Clearly, the six short measurements were individually plausible even

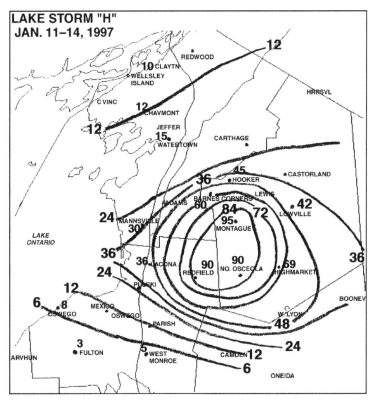

5.3. The Buffalo Weather Forecast Office's map of total snowfall from the multiday storm shows a peak 95-inch total at Montague.

though twenty-four hours of compaction would have produced a sum less than its parts.

Unfortunately, the region's Doppler radar station, then located at Griffiss Air Force Base, near Rome, was not operating at the time, and the radar beam from the next closest radar, at Binghamton, overshot the storm. Radar images might have shed light on the pattern of snowfall intensity. Although no connection is apparent, the air base was slated for closure, along with many other military installations, and the radar station was moved to Montague in 1998 because of the location's superior coverage.[19]

Imagery of another kind—cable television—accounted for much of the Montague measurement's notoriety.[20] Because a large lake-effect snowstorm was forecast, the Weather Channel had sent a crew to Buffalo to report on the storm, which unfolded dramatically, as predicted. Because timely reports phoned in by snow spotters were automatically included in announcements broadcast on NOAA Weather Radio and posted on the Buffalo forecast office's website, Weather Channel staffers who were monitoring developments on Tug Hill did the math and informed the forecast office that the six reports covering the period 1:30 p.m. Saturday (January 11) through 1:30 p.m. Sunday (January 12) added up to an unprecedented 77 inches. Local Weather Service personnel checked satellite data and reports from other spotters on Tug Hill, and concluded that a 77-inch twenty-four-hour snowfall was indeed plausible.

Meanwhile, the Weather Channel was already reporting the measurement as a new twenty-four-hour record, and other media, national and local, had picked up the story. "Pressured [for] a timely decision" on whether to accept the 77-inch measurement, Buffalo officials requested guidance from Weather Service Headquarters, which confirmed the Buffalo office's right to accept the Montague measurement as the new record, which it did on Wednesday morning (January 15). As the controversy evolved, Weather Service officials were criticized for letting the local office make the call. Not surprisingly, the committee recommended an official, more rigorous review process, implemented later that year when NOAA set up the National Climate Extremes Committee.[21]

The flawed measurement afforded an opportunity for committee members—Nolan Doesken certainly, and probably a few others—to complain of sloppy snowfall measurements throughout the country.[22] Some cooperative observers were not using snowboards, and even a few NWS Weather Forecast Offices were reporting daily and six-hour snowfall totals obtained by summing hourly measurements. Official guidelines called for taking snowfall measurements at least once a day but no more than four times in twenty-four hours, and no more than once every six hours. Observers were also required to measure depth of snow on the ground—a separate measurement useful for estimating both the snow load on roofs and the potential for flooding—but some were apparently subtracting yesterday's depth of snow from today's, a practice that "erroneously under-measur[ed] new snowfall."[23] The report called for better training and increased adherence to standardized procedures so that data collected for climatic studies would be suitably comparable.

A related issue is an increased need for truth-in-labeling if real-time reports from snow spotters are to coexist with the ostensibly more rigorous and complete reports of cooperative observers, whose more uniform network and broader range of measurements is an essential part of the climatologist's database.[24] Rather than denigrate the efforts of well-intended volunteers, Doesken and some of his Colorado colleagues formed the Community Collaborative Rain, Hail, and Snow Network (CoCoRaHS) to increase public understanding of weather science and to form a wider, denser network of enthusiastic amateurs committed to collecting precipitation data. The number of active volunteers in Colorado increased from 110 in 1998 to 1036 in 2004, and to over 1,500 in 2010, and there are now CoCoRaHS organizations in all fifty states.[25] In 2009 the Buffalo forecast office merged its spotter network into CoCoRaHS, which receives added encouragement and training from regional and county coordinators.[26]

Active CoCoRaHS observers measure precipitation once a day, between 6 and 9 a.m., using a simple rain gauge and a ruler for snow. In addition to giving forecasters (and each other) valuable information, they're also a recruiting pool for the older network of cooperative observers, who face stricter entrance requirements and closer supervision, including inspection of their stations and instruments once or twice a year. Nowadays Buffalo's CoCoRaHS and the area's less numerous "co-ops" share the same Snow Measurement Guidelines.[27]

However valuable to weather scientists and operational forecasters, systematic snowfall measurements for periods less than twenty-four hours were never part of the official record. After all, how many competent volunteer observers would willingly lose sleep measuring snow every half hour or hour during a multiday snowstorm? I was thus puzzled to discover "record" snowfalls for periods ranging from 20 minutes to 22 hours in Christopher Burt's *Extreme Weather: A Guide and Record Book*. A table titled "Selected Record Point Snowfalls" lists forty-four events for periods ranging up to twelve months and ranked from shortest to longest.[28] The table lacks a source note or other attribution, and Google turned up no similar listing online. That eight of Burt's ten shortest periods were not only in the United States but within seventy-five miles of where I live in New York State (fig. 5.4) triggered an urge to track down the unidentified sources of these obviously anecdotal measurements or at least confirm their plausibility.[29] Although NOAA's *Climatological Record* might not mention the 5.0 inches of snow that fell within 20 minutes at Turin, New York, on December 22, 1993, our local media would surely have reported the accompanying consequences of such an intense storm.

Indeed, two sources confirm the plausibility of the Turin event. Syracuse had morning and afternoon newspapers at the time, and both reported that an observer at the Snow Ridge ski resort, in Turin, had "recorded 5½ inches of snowfall during one 20-minute period."[30] Both newspapers also noted that the severe storm caused several accidents. State Police had to close Interstate 81 for several hours, and over 1,000 households lost electric power. The next day the resort's answering machine was telling callers that snow making was unnecessary because "our snow guns are buried."[31] Although Snow Ridge was no longer one of the National Weather Service's cooperative observers, NOAA's monthly publication *Storm Data* credited Turin with a 36-inch snowfall, the largest in the area between December 22 and 24.[32] Even so, a database search failed to find any mention of the intense 20-minute snowfall in the meteorological literature.

The next two records, observed over thirty minutes and one hour at Copenhagen, New York, were easy to verify. In response to a query about Burt's sources, Jessica Rennells, a climatologist at the Northeast Regional Climate Center, sent me an issue of *Storm Data*, which noted,

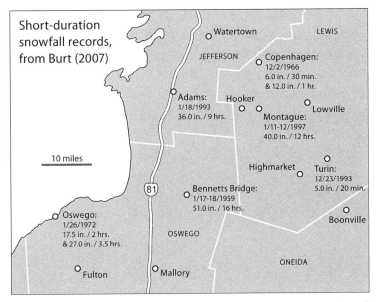

Short-duration snowfall records, from Burt (2007)

Watertown

LEWIS

JEFFERSON

Copenhagen:
12/2/1966
6.0 in. / 30 min.
& 12.0 in. / 1 hr.

Adams:
1/18/1993
36.0 in. / 9 hrs.

Hooker

Lowville

Montague:
1/11-12/1997
40.0 in. / 12 hrs.

10 miles

Highmarket

Turin:
12/23/1993
5.0 in. / 20 min.

81

Bennetts Bridge:
1/17-18/1959
51.0 in. / 16 hrs.

Oswego:
1/26/1972
17.5 in. / 2 hrs.
& 27.0 in. / 3.5 hrs.

Boonville

OSWEGO

ONEIDA

Fulton

Mallory

5.4. Snowfall records for periods less than twenty-four hours as listed in Christopher Burt's *Extreme Weather*.

in minimalist English, "Reliable reports indicate that snow fell at rate of 6 inches for 30 minutes and 12 inches per hour between 2 p.m. and 3 p.m. on 2nd at Copenhagen with visibility reduced to less than 50 feet."[33] The event was also highlighted in *Climatological Data*, in a monthly summary by state climatologist A. Boyd Pack, who wrote, "A competent observer at Copenhagen measured approximately 60 inches of new snow as the storm diminished late on December 2. Between 2 and 3 p.m. on the 2d snow fell at the rate of 6 inches per half hour and visibility was reduced to 15 yards, according to this same observer."[34] Copenhagen was not in the list of cooperative stations, and the "competent observer" remained anonymous.

Not so for Oswego, New York, recognized for two extreme measurements, both on January 26, 1972, when Robert Sykes, who had joined the meteorology faculty at the State University College at Oswego after retiring from the air force, measured 17.5 inches of snow in 2 hours and 27 inches in 3.5 hours during what he called a "snowburst" because its intensity and short duration reminded him of a cloudburst. As Sykes reported in *Weatherwise*, he had measured snow seven times between 12:45 and 4:14 p.m. (209 minutes) for two 15-minute periods followed by five 30-minute periods, which yielded a total of 24.6 inches during 180 minutes, which he expanded to 210 minutes (3.5 hours) by adding 2.4 inches as a "reasonable estimate during missing minutes between observation periods"—time needed to make and record his measurements and clear the snowboard under no doubt difficult conditions.[35] *Storm Data* and *Climatological Data* reported heavy snowfalls in Oswego County for January 25–27, including a "record 24-hour total of 37 inches at Mallory," but made no mention of Sykes's measurements, taken several miles away from the official cooperative observer at Oswego East.[36] Although Sykes didn't flaunt it, the 17.5-inches-in-2-hours record listed by Burt was apparently constructed by summing the snowfall for the four 30-minute periods between 1:58 and 4:14 p.m. while ignoring the intervening gaps of 4, 5, and 7 minutes.[37]

More mystifying is the 36.0-inch snowfall alleged to have occurred at Adams, New York, on January 18, 1993. *Storm Data* said nothing about heavy snow in the area on the eighteenth, or even the seventeenth or nineteenth, although *Climatological Data* reported 6 inches on the eighteenth at Hooker, only an inch at Bennetts Bridge, and just a trace at Oswego East and Watertown.[38] Even so, AccuWeather's Today in Weather History blog asserted that for January 18, 1993, "Thirty-six inches of snow fell in just nine hours at Adams, NY."[39] No source was listed, but I suspect the unnamed contributor gleaned the factoid from Burt's book. Google, which uncovered the AccuWeather posting, also turned up a similar note—"Adams was buried under 36 inches of snow in only 9 hours"—by blogger Charlie Wilson on Examiner.com.[40] Wilson, however, had assigned the storm to 1994, not 1993.

Pursuing this lead, I found several big snowfalls in the area on January 18, 1994, for which *Climatological Data* listed 10 inches at Bennetts Bridge, 9.4 inches at Highmarket, and 6 inches at Watertown.[41] But nothing for Adams, which had no official observer. Nevertheless, the *Syracuse Post-Standard* reported poor driving conditions between Adams and Watertown on January 18 and 19 in a story titled "Snow Buried North Counties: A Lake-Effect Storm Drops Three Feet on Parts of Oswego and Jefferson Counties," and in a summary of winter 1993–94 in *Weath-erwise*, three National Weather Service snow experts noted that in mid-January, "Adams, New York, got 60 inches in less than 24 hours."[42] But nothing about three feet in nine hours, which, if it happened at all, occurred in 1994.

Less problematic is Burt's seventh shortest extreme snowfall: the 40 inches that fell within 12 hours at Montague on January 11–12, 1997, date of that site's disputed 77-inch record. The longest continuous measurement in the contested 24-hour summation, it was accepted by the NOAA study team as a measurement but not as an official record.[43] Similarly vetted is the 51-inch, 16-hour snowfall at Bennetts Bridge on January 17–18, 1959, cited in the Special Weather Summary in that month's *Climatological Data*.[44] Sometimes hailed as the snowiest place in the state, Bennetts Bridge is part of the cooperative observer network.

While climate scientists reject records for periods shorter than twenty-four hours, longer periods formed by summing two or more twenty-four-hour totals yield a variety of official superlatives, including extremes for individual months and for the entire snow season. What's more, the official snowfall database lets climatologists not only calculate cumulative snowfall tallies for any day in an average season—useful for assessing whether the current year is running ahead of or behind what's "normal"—but also retrieve cumulative daily snowfall tallies for the most and least snowiest years on record.

For cities like Syracuse, where residents take pride in besting old records as well as upstate rivals like Albany, Buffalo, and Rochester, a graphic on the local newspaper's weather page (fig. 5.5) makes the running comparison with past records more interesting than the big-city snow race. By Christmas Day 2010, for instance, the city had registered a record December snowfall (over six feet), which raised the possibility of breaking additional records. Although a comparatively dry stretch between Christmas and New Year's dashed hopes that our snowiest December would also be the snowiest month ever, the graph shows that extreme annual totals typically reflect a few huge snowstorms rather than the steady, persistent accumulation most characteristic of lake-effect snow. Although Syracuse won the Golden Snowball Award for 2010–11 by a comfortable margin,

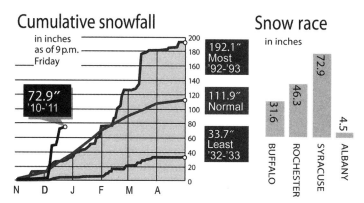

5.5. Replica of the *Syracuse Post-Standard*'s weather-page illustration comparing cumulative snowfall through Christmas Eve 2010 with snowfall received in a normal year as well as during the most and least snowiest years on record.

its final tally disappointed local snow boasters by falling 13.1 inches short of the record 192.1 inches measured in 1992–93. The following year Syracuse, with only 50.6 inches, lost to Rochester, with 59.9.

The National Climatic Data Center (NCDC), which maintains the country's Snow Climatology database, lets visitors to its website interrogate the snowfall and snow cover experience of individual stations.[45] Syracuse residents can learn, for instance, that the likelihood of a measurable snowfall rises from 0 percent for September, to 34.1 percent for October, 97.6 percent for November, to 100 percent for December through March, before declining to 85.5 percent for April, 10.6 percent for May, and 0 percent for June. The biggest one-day May snowfall on record is the 2.1 inches received on May 12, 1996, memorable locally because it coincided with both Mother's Day and the Syracuse University commencement, delayed a few hours while walkways were cleared.

For each state, the NCDC identifies locations with extreme conditions for ten different measures. Figure 5.6 shows that all of New York's extreme places are either on or adjacent to Tug Hill. Watertown's greatest daily snowfall, 45 inches measured for November 15, 1900, reflects an impressive 104 years without missing data. Three stations (Highmarket, Mallory,

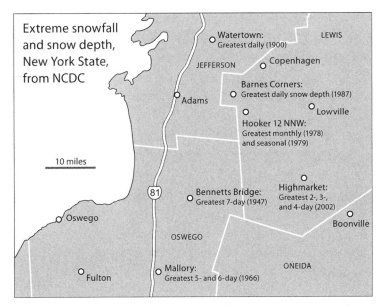

5.6. Locations of snowfall and snow depth extremes in New York.

and Bennetts Bridge) account for the six multiday extremes, representing two- to seven-day periods with snow every day. Hooker 12 NNW—the station's name indicates a measuring point twelve miles north-northwest of the hamlet of Hooker—registered the state's greatest seasonal and greatest monthly snowfalls, 379.5 and 182 inches, respectively. By contrast, Barnes Corners, with an 84-inch snow cover on January 25, 1987, claimed the record for greatest daily snow depth. While other places in the region might have experienced snowier days or seasons, these places made the list because their measurements reflect the higher standards of the cooperative observer network.

New York's official extremes map differs markedly from the corresponding map for Michigan, the only other Great Lakes state with comparable snowfall and snow depth extremes. As figure 5.7 shows, Michigan's extremes are dispersed broadly across the northern part of the state. On the rugged Keweenaw Peninsula, where some homes have a second-floor entrance,[46] Copper Harbor 3 WNW registered the state's greatest seasonal and monthly snowfalls, 263.5 and 129.5 inches, while Herman, about sixty miles south-southwest, measured the greatest daily snowfall, 30 inches. At the eastern end of the Upper Peninsula, DeTour Village, with a 70-inch snow cover on February 26, 1962, accounted for the greatest daily snow depth, and the Sault Ste. Marie Weather Service Office reported the greatest two-day snowfall. Farther south, on the Lower Peninsula, Petoskey captured Michigan's remaining multiday records, thanks to multilake snowbands crossing Lakes Superior and Michigan and the orographic effect of elevated hills at the northern end of the relatively flat peninsula. These latter five extremes reflect a single persistent lake-effect snowstorm during which cold air streaming southward from Canada dropped 85 inches on Petoskey over the seven-day period December 23–29, 2001—a total snowfall rivaling the seven feet dumped on Buffalo that week.[47]

Whether extreme or modest, reliable measurements are indispensable in estimating the likely weight of snow on roofs, a hazard largely ignored during most of the twentieth century, to the regret of many building owners in lake-effect areas. Buffalo's veteran Weather Bureau forecaster Bernie Wiggin introduced his classic 1950 *Weatherwise* article on the "Great Snows of the Great Lakes" by recounting a mid-October 1930 snowstorm

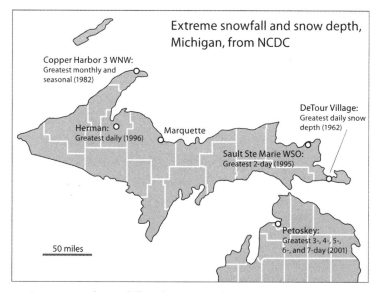

5.7. Locations of snowfall and snow depth extremes in Michigan.

that caved in a hotel roof in the village of Angola, south of the city, and brought down "hundreds of small buildings" along the Lake Erie shore with upwards of three feet of heavy, wet snow.[48] Snow collapse is also a serious concern at barn workshops run by the New York State Historic Preservation Office.[49] Even snow fighters are vulnerable, as demonstrated in January 2008 in Fulton, midway between Syracuse and Oswego, when a build-up of snow collapsed the roof of the public works garage at 5:20 a.m., trapping most of the city's snowplows, salt spreaders, and garbage trucks.[50] No one was hurt, and neighboring municipalities helped clear Fulton's streets.

Although the Great Lakes region is not Ground Zero for North American blizzards, the cumulative consequences of recurring snow can be costly and disruptive. In early 2010, for instance, a wall of an old brewery next to Interstate 81 in Syracuse collapsed under more than a foot of new snow.[51] The highway was closed to northbound traffic for 22 days, while city and state officials bickered over who would pay to demolish the dilapidated four-story building, which threatened to topple onto the roadway.[52]

Roof collapse under heavy snow is one of three types of structural failure for which building codes need maps. Severe winds and seismic shaking, which can be far more devastating, impose bursty but mostly horizontal forces, whereas snow accumulating on a roof exerts a growing downward pressure.[53] Mapping snow's impact requires the collaboration of structural engineers, who can design for an extreme snow load, and climatologists, who have the data and methods for estimating its geographic variation. Their maps portray what's called *ground snow load*, defined as the force per unit area of the snow accumulated on a horizontal surface and typically measured in pounds per square foot (*psf*). After the climatologist extrapolates a plausible extreme snow load from available data, the engineer or architect must consult the map when tailoring a design to a particular site and situation. Local terrain might encourage drifting, while the exposure, geometry, and thermal characteristics of a particular roof can create an imbalanced load if, for example, snow accumulates more heavily on the leeward side of the roof ridge or cascades from a higher roof to a lower one.

As in other attempts to map the plausible impacts of a natural hazard, the climatologists and engineers responsible for snow load maps bowed to the mystique of round numbers by specifying fifty years as the time span within which the minimum design load will be exceeded, on average, in only a single year. Fifty years was not much beyond the best records available when the enterprise began and is not so precise as to imply greater precision than warranted. Designing for a ten-year extreme event would invite too frequent failures whereas a hundred-year snow load standard might exaggerate the reliability of the model and add unnecessarily to construction cost. The actual design strength typically exceeds the official strength required by the building code because conscientious engineers and architects try to anticipate exceptionally severe weather or just plain bad luck. A safety factor saves lives and avoids lawsuits.

In the lingo of environmental cartography, fifty years is the *recurrence interval* associated with an *exceedance probability* of 0.02, which is another way of saying there is only a 2-in-100 (or 1-in-50) chance that the actual ground snow load will exceed the estimated snow load in any given year. Environmental cartographers are reluctant to talk of a "fifty-year snow load" because people who don't understand probabilities might think an extremely deep snow last year rules out an equally severe event for the next half century. When Mother Nature rolls the dice, next year's disaster could be even worse.

Probability-based building codes, which emerged in the late 1960s,[54] were largely an academic curiosity until 1980, when the National Bureau of Standards refined the procedure.[55] In 1982 the American National Standards Institute (ANSI) implemented this approach in a new standard that included risk maps for winds, earthquakes, and snow load.[56] The snow load map was provided by the Cold Regions Research and Engineering Laboratory (CRREL) of the Army Corps of Engineers, which oversees design and construction at military bases. CRREL became interested in mapping snow load in the mid-1960s, but little happened until the Bureau of Standards released its report in 1980.[57]

To make their map of 2 percent snow loads, CRREL scientists had to integrate a sparse network of 184 National Weather Service offices, which had measurements for both snow depth and snow weight—the water equivalent of snow on the ground—with the denser network of more than 9,000 co-op stations, with data only for snow depth.[58] Additional stations and a generally longer period of record led to a new map, published in 1995 by the American Society of Civil Engineers, which had assumed responsibility for the nation's structural design standard in the late 1980s.[59]

Standards are revised roughly every five years, after careful scrutiny by committees of experts, but many parts remain unchanged. CRREL's first snow load map, published as part of what was called the A58.1-1982 standard, was revised slightly (but only for Minnesota, Montana, and the Dakotas) for the ASCE 7-88 standard,[60] and this updated version was also used for ASCE 7-93. The second map, introduced in the ASCE 7-95 standard, also appeared in the 1998, 2002, 2005, 2007, and 2010 editions. Each revised standard is made available to the organizations that write the nation's primary building codes, which in turn are made available for adoption by states or localities, which can accept or reject the code or modify it to address local concerns.[61]

Lake-effect snowbelts are readily apparent on the first CRREL map. My slightly reduced excerpt (fig. 5.8), which focuses on the Great Lakes, reproduces the solid black fill used to show "areas [where] extreme local variations in snow loads preclude mapping at this scale."[62] That's right: uncertainty was so great that CRREL refused to put numbers on the map for a broad zone that included the Appalachians southeast of Lake Erie, the Catskills and the Adirondacks in New York State, and the mountains of western New England as well as the rugged Appalachians from Pennsylvania southward into western North Carolina. Vast stretches of the mountainous West are similarly blanked out. By contrast, gray shading represents a slightly greater certitude where mapmakers provided numbers but warned "the zoned value is not appropriate for certain geographic settings, such as high country." Architects and structural designers had to be wary of both orographic and lake-effect snow.

On the second CRREL map (fig. 5.9) gray zones overprinted with "CS" (for case study) carried a warning identical to that for the black areas on the earlier map.[63] Although the two maps differ noticeably in their zones of uncertainty, strong similarities are apparent, particularly for the lake-effect snowbelts along the eastern shore of Lake Michigan, the southeastern shore of Lake Erie, and the southeastern shore of Lake Ontario as well as eastward across Tug Hill into the Adirondacks. That some of the

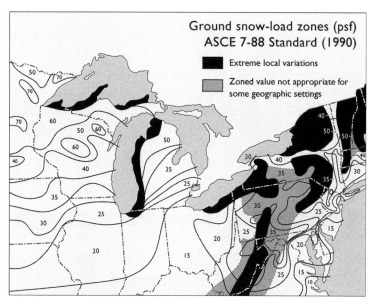

5.8. Ground snow loads in the Great Lakes region according to ASCE Standard 7-88, approved in 1988 and published in 1990, 1993, and 1994. Zones and boundaries within the area shown are identical to those for the ANSI A58.1-1982 standard.

blacked-out areas on the first map now have zoned snow loads no doubt reflects CRREL's increased confidence in probability modeling and its updated data set.[64]

A 2002 CRREL report outlined a straightforward strategy for local officials and builders who need a site-specific case study. Three staff scientists and a trio of structural engineers who volunteered their time established ground snow loads in each of New Hampshire's 259 towns.[65] Individual case studies averaged only thirty-two minutes, but as the authors recommended, "at least three people should independently do each case study." They also cautioned readers "never put all your faith in results from a single station" and "not [to] completely dismiss any station because its values do not fit well with others around it."

New York State was equally ripe for this kind of close scrutiny. The 1995 ACSE map, which put more than half of the state in a CS zone, was hardly better than its 1982 ANSI predecessor, which had painted an even

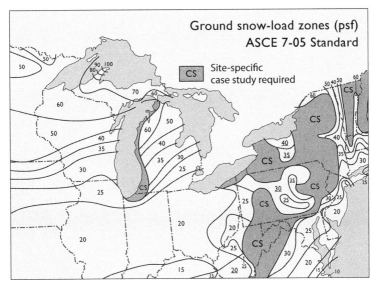

5.9. Ground snow loads in the Great Lakes region according to ASCE Standard 7-95, and reproduced unchanged in editions published in 1998, 2002, 2005, 2007, and 2010. Underlined numbers identify zones with a threshold elevation above which a site-specific case study is required.

larger area black or gray. Not compelled to adopt a building code aligned with an ANSI or ASCE standard, state officials addressed snow load with three different maps over the past three decades. Figure 5.10, which juxtaposes the three maps with the second CRREL map, shows that the New York Department of State, which administers statewide building codes, avoided the ambiguity of CS zones while recognizing relatively heavy snows on Tug Hill as well as in the Adirondacks and the Catskills.

The 1984 map (fig. 5.10, *upper left*) is plausible but crude. An accompanying table provided guidance in adjusting the mapped loads for roofs with a slope between 0° (flat) and 60°, but prominent peaks over Tug Hill and the Adirondacks suggest the zone boundaries were based on only a small number of data points.[66] According to an Albany official who was unable to identify the map's origin, it had been in use since the early 1950s, before probability-based building codes were the norm. The map's zones, he conceded, "may be best described as uncorrected roof snow load values."[67]

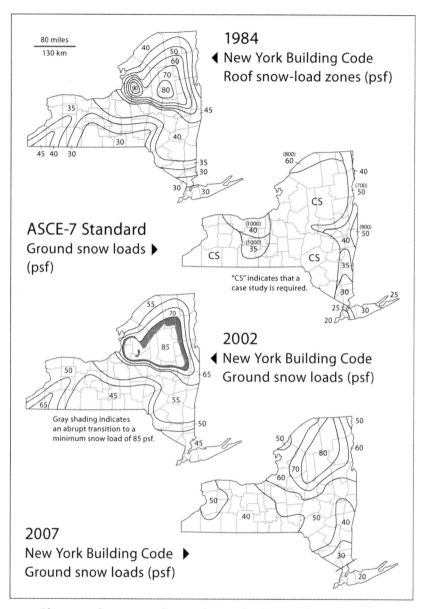

5.10. Changing depictions of ground snow load in building codes for New York State.

Two decades later New York was clearly aware of the national standard—sort of. Although the state building code published in 2002 states "design snow loads shall be determined in accordance with Section 7 of ASCE 7," the accompanying snow load map (fig. 5.10, *lower left*) revealed no reliance on CRREL cartography (fig. 5.10, *upper right*).[68] Five years later a new code introduced a very different map (fig. 5.10, *lower right*), the simplest of the three, which warned that "sites at elevations above 1,000 feet shall have their ground snow load increased from the mapped value by 2 psf for every 100 feet above 1,000 feet."[69] Outside the CS zones its minimum design loads are generally higher—and thus more cautious—than those on CRREL's 1995 map.

Architects eager for a second opinion might consult the Defense Department's 2007 report *Unified Facilities Criteria (UFC): Structural Load Data*, which lists design constraints for snow load and other hazards for a variety of locations, foreign and domestic, including ten sites in New York.[70] In listing slightly lower loads for Buffalo (45.1 psf), Syracuse (40.1 psf), Griffiss Air Force Base in Rome (59.9 psf), and Fort Drum (70.0 psf), the DoD report indirectly endorses the 2007 New York map as dependable if not slightly conservative.

Leery of cartographic risk assessment and the consequences of a bad call, the UFC report's authors added a table of safety factors for various "occupancy categories."[71] Published snow loads could be reduced by 20 percent for Category I, which includes agricultural facilities and other buildings "that represent a low hazard to human life in the event of failure." By contrast, a 50 percent increase was recommended for Category V, which includes power-generating stations, nuclear storage bunkers, and other "facilities designated as national strategic military assets." This pointed rejection of one-size-fits-all design loads underscores the difficulty of estimating snow accumulation.

Architects typically address extreme snow loads with either sturdy framing or a steep roof. Home builders in rural parts of the Great Lakes region have found metal roofs especially useful even when the pitch is as gentle as 3-in-12. A low coefficient of friction prevents a build-up of snow, and metal roof coverings can last much longer than the asphalt shingles widely used in residential construction. On the downside, snow tends to slide off a metal roof suddenly and en masse, making snow guards and diverters essential to protect vulnerable lower roofs and deflect snow avalanches away from entrances.[72] Because sliding snow can ruin foundation plantings, some homeowners cover arborvitae and other fragile shrubs with custom-made A-frame shelters (fig. 5.11).

In many cities and suburbs metal roofs carry a stigma, perpetuated perhaps by recollections of rusting corrugated roofs on dilapidated barns and abandoned farmhouses. Although a good metal roof can be quite pricey these days, especially when crafted in copper, an attractive baked-enamel roof is not only affordable but longer lasting than the extrathick, "architectural-grade" asphalt shingles with a "dimensional" texture designed to mimic a traditional slate roof better than the thinner, flatter asphalt shingles once in vogue.[73] Slate and copper offer durability and authenticity—more so than wood shingles, which rot or too easily catch fire—and enameled sheet metal is not only more durable than asphalt but arguably more authentic because it's not intended to resemble something

5.11. A-frame shrub shelters in Boonville, New York, on the southern fringe of Tug Hill.

else. Metal shingles that mimic slate better than asphalt are durable and expensive but hardly authentic.

Snow country homeowners in need of a new roof seem to agree, especially in rural areas where houses are farther apart and residents care less about what neighbors might think. In recent years a few metal replacement roofs have even appeared in the Syracuse suburbs where I live, but the trend is more noticeable up north toward Tug Hill. A systematic survey might eventually reveal a Metal Roof Belt, cartographically akin to the Corn Belt or the Bible Belt but stretching from the mountains of New England across the Great Lakes snowbelts—a shared strategy for coping with persistent lake-effect or synoptic snowfall.

6 Change

Well aware that a mild winter often follows a severe one, snow country residents sometimes ponder long-term trends. Have our winters become shorter, milder, and less snowy than in past decades, or has snowfall increased? Answers might be found in really good data, which are in sadly short supply, as well as in theories about climate change, confounded by the complex interactions of diverse physical processes and occasionally by political rhetoric at odds with systematic science. I'll stick my neck out—not very far, actually—by noting that both the evidence and the theory point to increased lake-effect snowfall in the Great Lakes basin. For snowfall in general, though, it's a different story.

Because the signal (as climate scientists call it) is clear but not blatant, I am reminded of a 1919 *Scientific Monthly* article by Harvard professor Robert DeCourcy Ward, who supervised Charles Brooks's pioneering doctoral dissertation. Convinced that "the only reliable evidence is that which rests upon instrumental records," Ward challenged the "widespread popular belief [in] 'old-fashioned New England winters'" by arguing that memories of "many heavy snowstorms [that made sleighing] possible for three or four months without a break" reflected a human tendency to recall, if not exaggerate, extreme events.[1] Anyone who grew up in open country but later moved to the city, where buildings are better heated and snow's impacts less severe, "naturally thinks that the winters are milder or less snowy than when he was a child."

Like a good scientist, Ward advocated a careful examination of data acquired with "accurate instruments, properly exposed and

carefully read [which] do not lie, do not forget, [and] are not prejudiced."
Even so, he was apparently willing to accept lack of evidence for a trend
as proof that no trend existed. "Scattered" records made it "difficult . . . to
draw general conclusions," he wrote, and "there is found no evidence of
any progressive change in the amount of snowfall." Much like an avowed
agnostic who's a closet atheist, he then concluded, quite generally, that
"weather and climate have not changed from the time of the landing of
the Pilgrims down to the present day." What he challenged, of course, was
the belief that winters were becoming milder.[2]

Ward's gratuitous certitude might be excused in part by a short, very
limited data set. In 1919 suitably complete, readily comparable data were
available for very few stations, mostly from the mid-1880s onward. Unless
one is prepared to trust anecdotal weather diaries—and I know of no scientist who is—trends in snowfall and snow days reaching back to the Pilgrims' landing are simply unknowable. Indeed, weather historian Robert
Ludlum (1910–1997), who documented extreme weather and its impacts
on colonial and nineteenth-century America in over a dozen books and
hundreds of articles, remained skeptical of all predictions, whether for
global warming or an impending ice age, and scrupulously ignored numerous opportunities to reflect on change, in one direction or the other, or
even to argue for no change, as Ward did.[3]

The weakness of the signal is confirmed by time-series graphs like
figure 6.1, which describes the annual snowfall record for Lowville, New
York, on the northeastern fringe of Tug Hill, roughly twenty-four miles
southeast of Watertown and thirty-six miles east of Lake Ontario. Observers for the cooperative volunteer network have collected snowfall data
here since November 1891, and these data have informed several recent
historical studies of lake-effect snow. Vertical bars representing total snowfall from August through July allow a ready comparison of individual snow
years. The graph reveals wide year-to-year variations as well as extreme
years like 1970–71, when total snowfall exceeded 250 inches. Lowville
also experienced exceptionally low years like 1982–83, when only 52.8
inches was reported.

Some of the graph's shorter bars (and some not so short ones as well)
reflect the observer's failure to enter a snowfall amount on his data sheet.

Snowfall in inches (August through July), 1891–2010

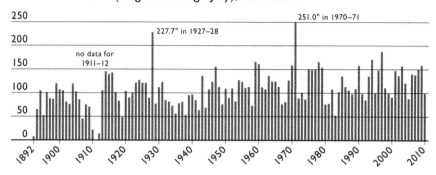

6.1. Vertical-bar graph of snowfall measured in inches at Lowville, New York. Each bar represents a period running from August through July and is aligned on the horizontal axis with the period's ending year.

The station was located at Lowville Academy (the local high school), and a stand-in observer might have either forgotten to measure snowfall or did not understand the guidelines. The leftmost bar, representing 1891–92, reflects only the 7.4 inches reported for November. For December 1891 through November 1892, the snowfall column on the monthly observers' sheets was left blank, even for days like December 16, 1891, when temperatures ranged between 17°F and 27°F, and the 0.57 inches of precipitation recorded most certainly fell as snow—perhaps a foot or more.[4] I'm also suspicious of the three snowfall measurements reported for November 1891: all three are exactly ten times the reported precipitation, as if the observer had melted the snow, measured its water equivalent, and then used the questionable 10:1 snow-to-water ratio to reconstruct the amount of snow. (It would be ironic if he had actually measured snowfall and then used the 10:1 ratio to calculate precipitation.) For December 1892, when snowfall reporting resumed, numbers rounded to the nearest inch and not in lockstep with precipitation amounts seem less likely to have been fudged.

When snowfall reporting seemed consistent, the National Climatic Data Center, in converting observers' sheets to an electronic format, coded blanks as zeros. Otherwise, blanks were treated as missing days, which were surprisingly common during Lowville's first few decades, as summarized in figure 6.2, which focuses on November through March,

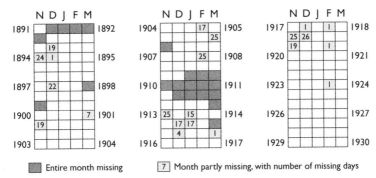

6.2. Missing data in the snowfall record for Lowville, New York, 1891 through 1930. Cells show the number of days for which no measurement was recorded during each of the five peak snow months, November through March.

the five snowiest months. Missing days were especially rampant from 1909 through 1912, with the longest break between January 1, 1910, and April 30, 1911. The observer, Charles J. Rice, apparently decided that snowfall was less important than melted snow. He was thirty-two at the time, had been taking measurements for several years, and was hardly in his dotage. His grandfather, Charles S. Rice (1827–1902), who set up the Lowville weather station in 1889, had recorded his first systematic snowfall measurements in November 1891—and for the next twelve months reported only melted snowfall. Go figure.

NCDC's quality control staff counted forty-one "breaks" in the Lowville snowfall record between 1891 and 2006, which seems a bit daunting insofar as each break is one or more consecutive months with missing days.[5] Although 16.2 percent of the daily values for these 115 years were missing, 77 percent of the months had complete snowfall data.[6] (Not wholly pessimistic, the NCDC tallies *complete* months as well as *missing* days.) Actually, Lowville's forty-one breaks are not unusual for a record this long. Watertown had fifty-seven breaks between 1894 and 2006, for which 15.2 percent of the daily values were missing and only 78 percent of all months were complete. At Hooker, closer to the center of Tug Hill, 44.2 percent of the daily values were missing for an even shorter record (1912–2006) with thirty-three breaks and only 51 percent of all months complete, while at Oswego East, on Lake Ontario, only 1.4 percent of daily values were missing for a still shorter period (1926–2006) with a mere seven breaks and 87 percent of months complete. For the same period, the volunteers at Boonville did even better, with a mere two breaks, only 0.8 percent of the daily snowfall values missing and 88 percent of all months complete. Surprisingly, volunteer observers occasionally surpassed their paid counterparts. At the Syracuse airport, where Weather Bureau personnel measured snowfall, there was only a single break between 1922 and 2006, but 2.9 percent of the daily values were missing and only 71 percent of the months were complete. In general, snowfall records were markedly better for the second half of the twentieth century than for the first.

Even so, Lowville's more recent data are hardly pristine. All snowfall data are missing for November and December 1971, and eleven years between 1950 and 2010 have one or more missing days. Because only thirty-nine years satisfied the NCDC's persnickety quality-control process used in determining the town's ten snowiest years, for a season running from fall through spring the official list ignores four years (1970–71, 1927–28, 1996–97, and 1993–94) even snowier than 1958–59, accorded first place with only 166 inches of snow.[7] And although sixty-four years were used to determine Lowville's top ten snowiest winters—meteorological winter is defined as the three months December through February— the NCDC's list of verified extremes omits eight years (from highest to lowest: 1970–71, 1977–78, 1927–28, 1993–94, 2008–9, 1996–97, 1989–90, and 2005–6) with more snow than 1978–79, the official snowiest winter.

Climate scientists who analyze snowfall data typically plot their numbers with dots rather than bars, which get in the way when they add lines describing trends. Figure 6.3 includes two trend lines, one straight and the other curved. The straight line provides a statistical "best fit" to the data.[8] Its gentle upward slope from left to right shows a slow, steady increase in seasonal snowfall between August 1924 and July 2010. A related probability value, which helps data analysts assess reliability, suggests that even though the points are widely scattered around the line, the relationship is hardly random.[9] By contrast, the smooth, curved line provides a more flexible generalization of the overall pattern, which is characterized by decreasing snowfall through the late 1930s, a steady increase from the

Snowfall in inches (August through July), 1924–2010

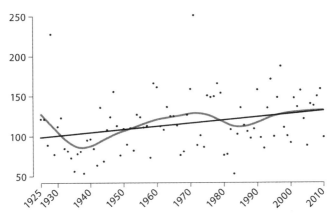

6.3. Dot plot showing snowfall at Lowville, New York, over the 86 August-through-July seasons ending in 2010. Points are aligned on the horizontal axis with the season's ending year. The straight black line, a statistical best-fit to the point cloud, shows the overall, long-term trend in the data, while the curved gray line reflects several shorter trends.

1940s through the 1970s, a short downturn from the mid-1970s through the mid-1980s, and a gradual increase since the late 1980s.[10]

While the overall trend is upward, the succession of decreasing and increasing phases shows how the trend varies with time period. Figure 6.4 underscores this point as well as the influence of extreme values. The long straight line extending over the entire period was calculated using all but the two most extreme snowfalls, 227.7 inches in 1927–28 and 251 inches in 1970–71. Removing these outliers moves the trend line slightly lower at the left and provides a marked improvement in the goodness-of-fit.[11] Another experiment, which retains the outliers but looks at four overlapping intervals, yields downward-sloping straight lines reflecting declining trends for two shorter periods, August 1924 through July 1944 and August 1958 through July 1983.[12] In both cases the data points are widely dispersed around the trend line, and neither of these downward trends is deemed statistically significant. By contrast, the two upward trends (for

August 1934 through July 1979 and August 1979 through July 2010) show less scatter around their trend lines, and both achieve statistical significance.[13] These results show how a biased analyst might spin the results by selecting a short time interval that produces the desired trend. They also suggest that long-term trends, based on more data points, tend to be more numerically meaningful than short-term trends.

Although the overall trend in figure 6.3 meets conventional criteria for statistical significance, it is reasonable to question the meteorological significance of a trend line with a slope of 0.389 inches/year. Although this works out to less than four inches over ten years, if the trend were to hold steady over a hundred years, Lowville's average annual snowfall would increase by more than three feet—certainly a significant amount.

How well does Lowville's trend represent Tug Hill and the wider Great Lakes region? A ready answer can be found in a study published in 2003 by geographic climatologist Adam Burnett and three collaborators, who compared snowfall trends at Lowville and fourteen other stations near the Great Lakes with ten stations outside the lake-effect region.[14] Their

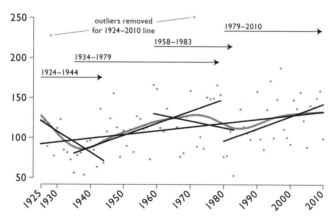

6.4. Four straight lines show best-fit linear trends for four comparatively short, slightly overlapping time intervals, and the long straight line covering the entire period is a best-fit trend line calculated with the two outliers excluded. The curved gray line is from figure 6.3.

lake-effect sites are mostly east of Lake Ontario or in upper Michigan, and the non-lake-effect sites are largely within 300 miles of the Atlantic Ocean (fig. 6.5). Slightly more than half of the stations had a sufficiently long, relatively reliable record extending from October 1931 through April 2001—the researchers felt they could safely ignore May through September. For the remaining stations, they used data for 1950 through 2001. Linear trend analysis similar to the procedure I used for Lowville (figs. 6.3 and 6.4) revealed a significant upward trend at eleven of the lake-effect stations (including Lowville), a nonsignificant upward trend at three others, and a nonsignificant downward trend at just one station, Boonville. By contrast, none of the trends at the non-lake-effect sites were statistically significant.

Lowville was one of four stations discussed in greater detail in another study of lake-effect trends published a year later.[15] Although geographers Andrew Ellis and Jennifer Johnson focused on change in snowfall over

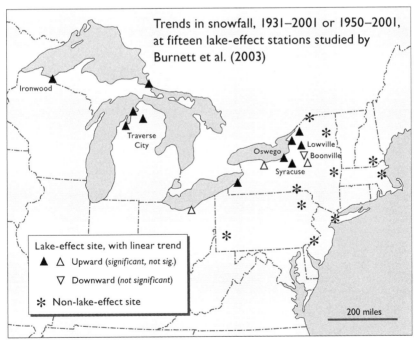

6.5. Linear trends in snowfall at the fifteen lake-effect sites and ten non-lake-effect sites studied by Adam Burnett and his colleagues. Trends at twelve of the lake-effect sites were both statistically significant and upward.

the forty-year period 1932–33 to 1971–72, they also looked at longer periods with available data, concentrated on the three winter months when lake-effect snow is most common (December through February), and considered not only the amount of snowfall but also two other measures: snowfall frequency, that is, the number of days with snow reported, and "snowfall event intensity," calculated by dividing the snowfall in inches by the number of days with snow. Interested in whether more snow meant less rain, they also examined trends in both total precipitation and the amount of precipitation falling as snow.

Lowville was selected for more detailed study because of its long record, which revealed a significant increase in snowfall from the late 1920s through the end of the century. The other three stations chosen for close scrutiny—Fredonia, in western New York, and Ironwood and Traverse City, in Michigan—also had comparably long records and, like Lowville, had been shown by Burnett and his coauthors to have experienced significantly increased snowfall between 1931–32 and 2000–2001 (fig. 6.5). At Lowville and the two western Great Lakes stations, snowfall events increased in number but declined in intensity, in contrast to Fredonia, where frequency decreased while intensity increased. Although trends at the four stations were hardly identical, increased snowfall was not matched by increased total precipitation during winter, suggesting that more snow generally meant less rain.

Concerned about data quality, Ellis and Johnson looked for the possible effects of changes in station location and the time of day when observers took measurements.[16] At Lowville, for instance, the observation site was relocated twice between 1932 and 1972: a move preceding the 1968–69 snow season raised the measurement site's elevation 11 meters (36 feet) and a move before the 1971–72 season lowered it 25 meters (82 feet). In addition, snowfall was measured in the morning through 1947–48 but in the afternoon thereafter. None of these changes had notable effects on the trends found at Lowville, and similar shifts at the other three stations were deemed inconsequential.

To assess broader patterns Ellis and Johnson drew on data for 568 climate stations to map trends for the region south and east of Lakes Erie and Ontario for the forty-year period 1932–33 through 1971–72. Areas to the lee of Lakes Michigan and Superior were omitted because too few stations had adequate data, and the area immediately downwind from Lake Huron was excluded because Canadian and US data were not readily comparable. After eliminating stations with incomplete or unreliable data, they estimated linear trends at the intersections of meridians and parallels one degree apart and plotted isolines showing the trends for several precipitation variables. Figure 6.6, based on two of their plots, shows a broadly increasing trend for snowfall (*left*) in contrast to an mixed pattern for total precipitation (*right*), which trended upward in three areas, most notably northeast Ohio and central New York (including Tug Hill) but generally downward. Note that the trend in total precipitation is scaled in millimeters per year, whereas the trend in snowfall is given in centimeters per year. Within the lake-effect region more snow meant less rain and perhaps less total precipitation. Other maps reinforced the conclusion that the lake effect accounted for more frequent, more intense, and more productive snowfall.

Trends in snow days also drew the attention of Daria Scott and Dale Kaiser, who focused on the period 1948–49 through 2000–2001 but

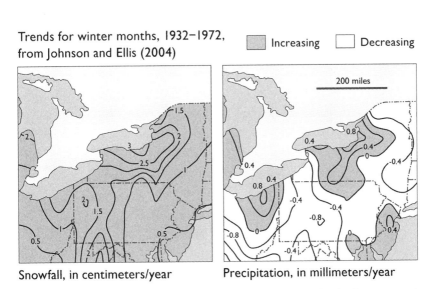

6.6. Trends in snowfall and total precipitation east and south of Lakes Erie and Ontario. Maps refer to the winter months (December, January, and February) between 1932–33 and 1971–72.

covered the entire conterminous United States.[17] They examined a snow season extending from October through May and defined "snow day" as any day with a measurable snowfall greater than 0.1 inch, the National Weather Service threshold for a mere "trace." Excluding stations south of 35° N and screening for data quality reduced the data set from 1,062 to 217 sites. The statistical procedure for estimating direction and strength of trend at individual stations used the conventional "95 percent confidence level" to assess statistical significance. Maps of the resulting trends showed the most significant change in the Pacific Northwest, where total snowfall, the number of snow days, and snow's share of total precipitation all declined.

Less dramatic but nonetheless noteworthy trends were apparent to the lee of Lake Ontario, where both snowfall and snow days increased. Figure 6.7, a black-and-white adaptation of color symbols on portions of two of their maps, required careful concentration. It shows consistent increases in western and central New York, with the increase in number of snow days (*upper map*) extending eastward into northern Vermont. Although the small number of sites in the snowbelts east of Lakes Erie and Ontario all showed more snow days and increased snowfall, these trends were not always statistically significant. Scott and Kaiser also found a longer snowfall season in lake-effect areas east of Lake Ontario.[18]

Numerous studies have noted similar trends. The earliest was published in 1960 by Weather Bureau scientist Jerome Namias (1910–1997), who looked at the influence of atmospheric circulation on snowfall at thirty-two stations throughout the eastern United States over the thirty-year period 1929 through 1959.[19] Although he mentioned the Great Lakes only in passing, his observation that "snow-storms are capricious and occur only a small number of days" during the three winter months underscores the importance of vast cells of high pressure, which can block or steer winter storms. Ten years later Val Eichenlaub, who popularized the term *snowbelt*, mapped the results of a statistical analysis of trends at eighty-two sites to demonstrate generally increased snowfall between 1920 and 1963 for sites immediately to the lee of the five Great Lakes.[20] Unlike Namias, he focused on a narrower area and emphasized the importance of the lakes as the source of moisture.

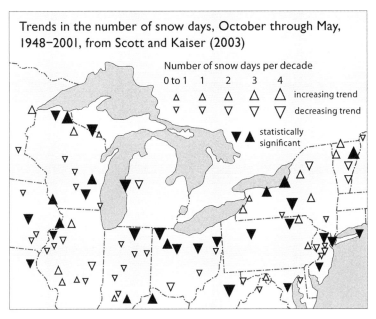

6.7. Trends in number of snow days (*upper*) and inches of snowfall (*lower*) as estimated by Daria Scott and Dale Kaiser for October through May between 1948–49 and 2000–2001.

Increased interest in climate change during the 1990s fostered renewed interest in snowfall trends near the Great Lakes. David Norton and Stanley Bolsenga, researchers at NOAA's Great Lakes Environmental Research Laboratory, in Ann Arbor, Michigan, drew upon US and Canadian data for more than 1,240 sites to create high-resolution gridded data sets for individual months from October 1950 through May 1980.[21] Despite marked variation within the region, analysis revealed a distinct increase in lake-effect snow but not in synoptic snowfall produced by wider, less localized winter storms. What's more, this upward trend in lake-effect snow mostly reflected increases during the middle and late parts of the snow season, from December through March, which implicated cooler winter and spring air temperatures. In a similar vein, University of Delaware researchers Daniel Leathers and Andrew Ellis, who looked at underlying atmospheric patterns, observed that conditions associated with lake-effect snow had increased from 1950 through 1982, while conditions associated with synoptic-scale snowstorms had decreased.[22] They attributed increased snowfall to increases in both the frequency and the intensity of lake-effect storms and suggested that warmer lakes with less ice cover might be responsible.

More recently a multi-institution team of snow researchers led by Kenneth Kunkel of the Illinois State Water Survey suggested that upward trends in lake-effect snowfall might be only half as large as previously thought.[23] And while increasing snowfall to the lee of Lakes Michigan and Superior seemed generally consistent and stable, trends for the snowbelts attached to Lakes Erie and Ontario were "mixed, depending on the period of analysis." And the trends for the two upper Great Lakes included increases in both snowfall and the snow's water content, presumably the result of warmer lakes and less ice cover, but the trend in water content was not statistically significant for the two lower lakes. Aware of the limitations of their sparse data set—a mere nineteen stations, all in the United States—the authors conceded the need for "a more comprehensive study" if climatologists were "to definitely determine cause and effect."

A truly comprehensive study must deal with the thorny trade-off between number of stations and data quality: improve one and the other deteriorates. Committed to basing their study on stations with homogeneous data for a meaningfully long period, the seven members of the Kunkel team had looked only at stations with thirty or more years of data located in areas known to receive substantial amounts of lake-effect snow. Each team member had independently evaluated each candidate station for homogeneity, comparing its data with records from the fourteen closest stations in order to weed out sites with anomalous, spatially incoherent fluctuations that might reflect inconsistent or biased measurements.[24] Their finicky screening threw out five of the seven longer-record stations studied by the Burnett group and two of the four stations accorded close scrutiny by Ellis and Johnson. Despite the problems noted earlier, Lowville, New York, survived the cut.

An explanation for lake-effect snowfall's long-term trends and quirky mood swings might lie in a teleconnection (literally, a long-distant link) to changing ocean currents or weather patterns thousands of miles away. The best known teleconnection involves El Niño, a noticeable warming of waters off the west coast of South America that occurs every few years around Christmas—El Niño is Spanish for "the child," more specifically the Christ Child. A year or so later an El Niño might be followed by much-cooler-than-normal sea surface temperatures, a phase called La Niña, Spanish for "little girl." As opposite, warm and cold anomalies in sea temperature, El Niño and La Niña are related to the Southern Oscillation, a corresponding shift in atmospheric pressure. Climatologists refer to the joint phenomenon as ENSO, the acronym for El Niño/Southern Oscillation.

As shown in figure 6.8, during winter a moderate-to-strong El Niño affects North American weather by fostering low pressure over the northern Pacific, which splits the jet stream and brings warmer-than-normal temperatures to the northern states, comparatively wet weather to southern California, and cool, wet weather to the Southeast. In an El Niño year, snowfall in the Northeast is generally lower than average.[25] A La Niña winter, by contrast, means high pressure over the northern Pacific, which forces the polar jet stream farther north and then southward across western Canada, resulting in generally warmer, wetter weather in the Northeast, but with no consistent or prominent effect on either regional or lake-effect snowfall.

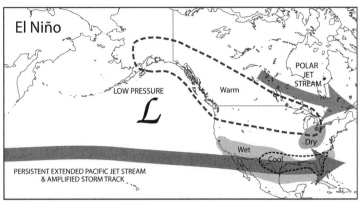

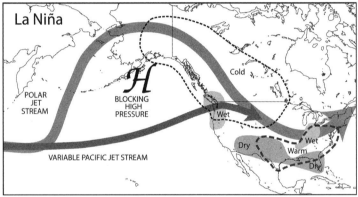

6.8. Weather anomalies and patterns of atmospheric circulation common January through March during moderate-to-strong El Niño (*upper*) and La Niña (*lower*) events.

Despite little apparent influence on the annual amount of lake-effect snowfall, ENSO can be useful to forecasters, especially during an El Niño year.[26] ENSO events can last a year or longer, and NOAA climatologists track not only their occurrence but also their strength, measured by the Oceanic Niño Index (ONI), based on shifts in sea surface temperatures in the central Pacific near the Equator,[27] and the Southern Oscillation Index (SOI), computed from changes in the difference in air pressure between Tahiti and Darwin, Australia.[28] For both indexes, a relatively high positive value reflects a strong El Niño, and a relatively large negative number indicates a strong La Niña.

An apparent lack of correlation between El Niño and Great Lakes snowfall might reflect researchers' treatment of winter as a homogeneous season. SUNY meteorology professor Richard Grimaldi detected distinct differences between early winter and midwinter during El Niño episodes—ONI and SOI show twelve El Niño years and eleven La Niña years between 1940 and 2007.[29] Although he examined data for only one station, Syracuse, New York, statistical analysis suggested that the polar and subtropical jet streams orchestrate the typical El Niño winter's slow start followed by heavier snowfall. During an El Niño year snowfall was generally lower than normal between December 21 and January 21, when above-normal temperatures often meant rain rather than snow, but noticeably higher between January 22 and February 22, when colder-than-normal temperatures and a warm Lake Ontario favored snow.[30] Even so, heavy snow in early winter was far more likely during an El Niño. By contrast, La Niña years were comparatively unremarkable.

Other teleconnections might be at work. In 2007 Daria Kluver (formerly Daria Scott), who studied North American snowfall patterns between 1949 and 1999, found generally increased snowfall in the Great Lakes region, particularly for December, January, and April, and a noteworthy negative correlation during October with the North Atlantic Oscillation (NAO), measured as the pressure difference between Iceland and the Azores.[31] A year earlier University of Wisconsin climatologist Michael Notaro and two co-authors reported that lake-effect snowfall around the Great Lakes was particularly strong when a negative NAO episode coincided with the positive phase of the Pacific–North American teleconnection pattern, involving pressure anomalies across much of the Northern Hemisphere.[32]

Recent interest in teleconnection underscores the importance of wind direction in the lake-effect recipe. As scientists who warn of global warning and its impacts are quick to point out, climate change is not just about average air temperature. Even without the more extreme disruptions—imagine the melting of the Greenland and Antarctic ice sheets, an accelerated worldwide rise in sea level, and the rerouting of the Gulf Stream—subtle, more local shifts in air circulation seem likely, and these shifts will surely affect weather patterns. In particular, an increased movement of polar air southward across the Great Lakes during winter would mean increases in both the frequency and the amount of lake-effect snow. That's probably what's happening now, according to Adam Burnett, who sees increases in lake-effect snow during the twentieth century as "a regional response to global warming."[33] And because regional trends in air circulation no doubt encompass smaller-scale shifts that affect the upper and lower Great Lakes differently, changing patterns in airflow might account for the aforementioned differences in snowfall trends between western Michigan and Tug Hill. Snowfall to the lee of Lakes Erie and Ontario would also include moisture from Lakes Superior and Huron—the multilake effect.

I've known Adam for over ten years and am fascinated with his research. He's a professor of geography at Colgate University who specializes in climatology and geographic information systems, and he has collaborated with several geologists in an ongoing study of lake sediments in Central New York. Sediment cores from the bottoms of deep lakes contain a signature of lake-effect snow that stretches back thousands of years and reflects salient shifts in the polar jet stream, which manipulates winds and weather patterns.[34] He has seen some of lake snow's widest mood swings. Adam's hunch is that lake-effect snow will increase for a while but eventually decline as warming winter temperatures become, as he puts it, "too warm for snow."

Likely consequences for Central New York are shrouded in the uncertainty common to all serious warming forecasts—*if* is always an easier issue for earth scientists than *when*. Computer models of global climate change predict that increasingly warm winters will, as Adam suggests, significantly stifle snowfall by the end of the century. University of Michigan climatologists Peter Sousounis and George Albercook, who examined the consequences of multiple scenarios for different parts of the Great Lakes region, noted large differences between two highly respected general circulation models, one forecasting warmer, much wetter conditions and the other suggesting a future that's only somewhat wetter but much warmer.[35] A related study by Kenneth Kunkel and colleagues at the Illinois State Water Survey looked at situations favoring heavy lake-effect snow—very cold air streaming across a large expanse of relatively warm open water—and noted a predicted reduction in these favorable conditions from an average of fifteen events per decade for the 1960s through the 1980s to an average of only seven events per decade for final thirty years of the current century.[36] Within the region, though, warming temperatures, more favorable to rain than snow, would cut the amount of lake-effect snow in half in the more southerly snowbelts southeast of Lake Michigan and south of Lake Erie but would have relatively little effect to the lee of Lake Superior.

For the short run, though, global warming does not preclude severe winters like 1976–77, when repeated assaults of Arctic air with a northwesterly flow brought record snow to areas downwind of the relatively warm Great Lakes.[37] As often happens, the disruption that year was far from even. Oswego registered between 75 and 100 inches of snow in a single storm, while Syracuse, which endured 42 consecutive snow days, December 31 through February 9, escaped a major event.[38] Western New York was less fortunate: a late January blizzard—driven largely by the passage of a cold front and thus more synoptic than lake effect—killed twenty-nine people, including nine found dead in cars trapped in snow.[39] It is clear that emergency managers cannot dismiss the threat of severe winter weather.

In the long run climate change will have profound impacts—some beneficial, others not—on government services, local economies, outdoor recreation, and the private snow-removal business. States and municipalities will spend less for snow plowing and road salt, homeowners will require less heating fuel, and farmers will have longer growing seasons, but without timely moisture from the spring snowmelt. A key concern is the amount and timing of rainfall: intense, more frequent thunderstorms will promote soil erosion rather than replenish soil moisture, and increased runoff means increased flooding. Moreover, substantially diminished snowfall could snuff out winter recreation activities like snowmobiling and skiing, while warmer year-round temperatures will foster diseases, insect pests, and invasive plant species rampant farther south. As much as climate scientists would rather not talk about winners and losers, some individuals in some regions will no doubt benefit, for a while at least, but claims that gains will outweigh losses or that net losses will be inconsequential are cynical if not fraudulent.[40] The broader, far-reaching impacts of a worldwide economic collapse and the geopolitical consequences of massive, sudden international migration would hardly be offset by the pleasures of year-round golfing near the lower Great Lakes.

Migration of another sort dramatizes the projected regional impacts of global warming. The Union of Concerned Scientists, a credible source of information on climate change, used "migrating climates" to underscore likely impacts on the Great Lakes region.[41] Separate maps for eight states and southern Ontario compare current seasonal average temperature and precipitation with conditions forecast for 2030 and 2095, when continued emissions of greenhouse gases will have made climates warmer and possibly wetter. Because average emissions scenarios predict Michigan will have much warmer and perhaps dryer summers and warmer, slightly moister winters, these projections can be described cartographically by recentering the state over northeastern Arkansas for summer and over Ohio for winter (fig. 6.9). A similar comparison would recenter New York over southern

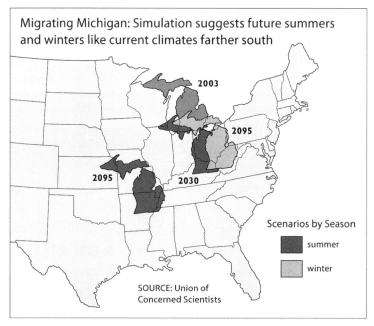

6.9. Simulated increases in temperature under the assumption of increasing emissions suggest that Michigan's climate will progressively match current conditions farther south for both seasons but markedly southwestward in summer while a bit eastward in winter.

Illinois for summer and over southeastern Pennsylvania for winter (fig. 6.10). Although the maps for both states show a more dramatic shift for summer, warmer winters would surely bring less snow by 2095. Of course, the notion of a state having a single average climate that migrates to another average elsewhere is a bit ludicrous, given Michigan's wide range of latitude and New York's contrast between Tug Hill and Long Island. Some enterprising cartographer will, I trust, devise a map that accommodates this intraregional diversity. After all, a more faithful depiction of Michigan's migrating climates must, at the very least, highlight marked differences between the Upper Peninsula and the rest of the state.

Although I'm fascinated with migrating climates as persuasive cartography, I'm a pessimist about public recognition of climate change as a threat worth addressing. Though the precise rate of warming seems fuzzy, the scientific case was settled years ago, however much a few scientists

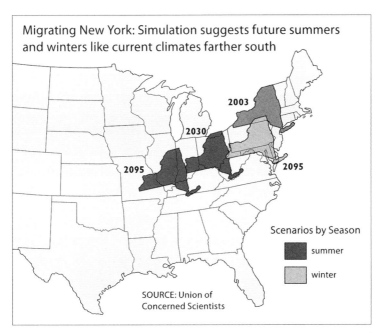

6.10. Simulated increases in temperature under the assumption of increasing emissions suggest that New York's climate will progressively match current conditions farther south for both seasons but markedly southwestward in summer.

funded by the petroleum industry say it isn't. But I'm skeptical the American public will ever get excited about the dangers, which seem too remote to people living far from the coast or the semi-arid Plains, and I doubt that even a better informed, more highly motivated American electorate can reverse disruptions that seem likely to grow steadily worse, even after the world's supply of fossil fuels runs out. The media's old-fashioned notion of journalistic balance, by giving undue acceptance to self-identified skeptics with PhDs, has made it easy for nonscientists to ignore solid analysis of systematically collected evidence when it is economically or politically inconvenient. Ironically, many of the recommendations skeptics resist would benefit cost-conscious homeowners as well as promote national security.[42]

7 Place

Ask a geographer if *location* and *place* are synonyms, and you might get a short lecture on geometry and imagination. Location, you'd learn, refers largely to latitude and longitude, or to relative position, as extolled in the mantra, "location, location, location." By contrast, place is more a matter of memory and meaning, as embodied in names like Gettysburg and Down East Maine. Some places are defined by a single event like the bloody Civil War battle that catapulted an undistinguished small Pennsylvania town into history books and travel guides. Others draw on a mix of traits like Maine's irregular coastline, picturesque waterfronts, and foggy mornings. For states bordering the Great Lakes, unusual winter weather makes the Upper Peninsula, Tug Hill, and the cities of Oswego and Syracuse (among others) distinctive places. Anyone who ever lived there during winter readily recalls lake-effect snow.

What makes lake-effect snow distinctive is not so much its depth as its frequency. Although the Great Lakes snowbelts occasionally receive a foot or more of snow in a single "event," the most distinctive feature is a pattern of repeated snowfalls, often just one or two inches a day, but enough to require shoveling walkways and plowing streets and driveways. To summarize this repetitiveness, I compiled figure 7.1 from snowfall records for January and February in Syracuse, for the dozen years 2000 through 2011. My graphic consists of 711 dots, one for each day, arranged by month and day in rows representing years. Each white dot (open circle) represents a day without measurable snow, each gray dot signifies a snowfall between 1/10 and 9/10 of an inch, and each black dot represents an inch or more of snow. The diagram's 198 gray and 236 black dots, which collectively outnumber the 277 white dots, show that 61 percent of the 711 days had more than a trace of snow, and 33 percent had an inch or more. What's more, the long runs of snowy days are as distinctive as their frequency. Most extreme was January 2009, with 17 continuous days with snow from the sixth through the twenty-second.

Equally distinctive are the year-to-year variations, which make the decision of how to pay for driveway snow removal a gamble. Landscapers in the snow removal business offer two rates: seasonal and per-plowing. Years like 2008 give them a modest extra profit because customers choosing the seasonal rate pay for plowings they don't need, while years like snowy 2003 give lump-sum subscribers their money's worth. It pays to read the contract carefully, though. Snow plowing services need to protect themselves against spiking gasoline prices as well as winters like 2010–11, when I paid $150 extra because the $395 contract price covered only 26 plowings. It's money well spent because shoveling out a driveway flanked by retaining walls is time-consuming and stressful. And the ambulance ride alone costs more than $400.

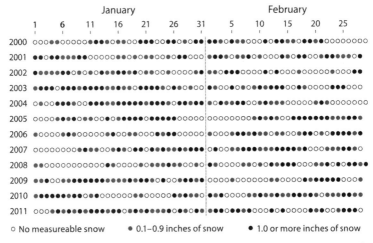

7.1. Days during January and February, 2000 through 2011, with measurable snowfall at Syracuse, New York. Black dots indicate one inch or more of new snow, and gray dots at least 0.1 inch.

Recurrent snowfalls that keep "lake-effect" on the lips of TV weathercasters in snowbelt areas can add up to impressive accumulations, both on the ground and in the climatological record. Snow cover is understandably longer lasting, on average, to the lee of the Great Lakes than along the Atlantic Coast, which is, as I've shown, more prone to large, infrequent, and occasionally memorable snowfalls delivered by synoptic storms fueled by the ocean. What's less likely in Baltimore, Philadelphia, and New York City is the perverse persistence of snowy weather represented by the long strings of black and gray dots in figure 7.1. It's no surprise that the snowbelts stand out when the National Climatic Data Center (NCDC) maps the mean annual number of days with an inch or more of snow (fig. 7.2). In an average year, the Great Lakes snowbelts, along with parts of the Appalachians and northern New England, have thirty or more days with a noteworthy snowfall—more than three times as many as most coastal areas. The Great Plains, home to most of the nation's blizzards, also has comparatively few snow days.

Another distinctive feature of the snowbelt climate is gloomy skies, especially in winter. This effect is apparent on the NCDC's map of cloudy

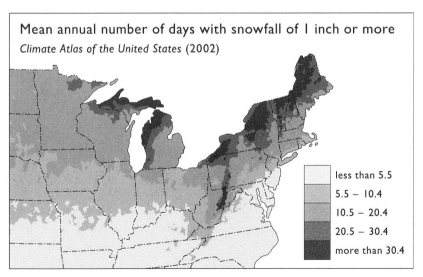

7.2. Average number of days with a snowfall of an inch or more, as mapped by the National Climatic Data Center for the period 1961–90.

days in January (fig. 7.3), which shows a wedge of frequent cloud cover extending southward from the Great Lakes into the Appalachians. Even when cold polar air is not moving moisture from the warmer lakes onto land as snow, our skies are often overcast. Throughout the winter, though, clouds vary in size, shape, and persistence. When the moisture is confined to a single long, thin (but not stationary) snowband, you can be shrouded in snow flurries one minute and have clear blue skies the next.

I've amused students and colleagues by printing a flyer on gray paper to advertise the "Annual Syracuse Cloud Festival," which is yearlong—January 1 to December 31 (fig. 7.4). It's an exaggeration, of course, yet what better way to cope with overcast skies than to celebrate them as an opportunity to read, catch up on chores, work out in a well-lighted gym, or think great thoughts? People prone to depression might blame their winter blahs on the weather, but I doubt they would be much happier in the Upper Midwest, where blue skies and bitter-cold temperatures are the norm in winter. In the Great Lake snowbelts, cloud cover acts like a blanket, to stem the loss of heat to outer space. Are cloudy skies any more depressing than frigid temperatures and horrendous heating bills?

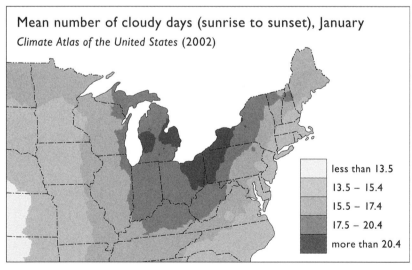

Mean number of cloudy days (sunrise to sunset), January
Climate Atlas of the United States (2002)

less than 13.5
13.5 – 15.4
15.5 – 17.4
17.5 – 20.4
more than 20.4

7.3. Average number of January days deemed cloudy, sunrise to sunset, by the National Climatic Data Center for the period 1961–90.

2011 ANNUAL

Syracuse
Cloud Festival

JANUARY 1ST – DECEMBER 31ST

7.4. Cloud Festival flyer, to celebrate overcast skies.

If overcast skies can stimulate creative thought, western New York's weather might have been the catalyst for Mormonism, Spiritualism, and women's rights. In the early nineteenth century, the area south of Lake Ontario, still very much on the frontier, witnessed numerous evangelical revivals in anticipation of the Second Coming of Christ and hundreds of conversions to nonconformist sects as well as more conventional Protestant religions.[1] In 1816 ten-year-old Joseph Smith, Jr., who founded the Church of Jesus Christ of Latter Day Saints, moved with his family to Palmyra, a village twenty miles southeast of Rochester. Visions that began at age fourteen culminated a few years later in a meeting with the angel Moroni, who led Smith to the Golden Plates, from which he translated the *Book of Mormon*.[2] The opening of the Erie Canal in 1825 brought a more cosmopolitan air to western and central New York, which provided a seed bed for other innovative faiths, most notably the Millerites, who anticipated the Second Coming on October 22, 1844, and disbanded after resetting the date a few times; the Shakers, a communal group that eschewed sexual intercourse and ultimately died out; the Oneida Community, known for its clever form of group marriage and successful silverware business; and the loosely organized Spiritualists, who communicated with the dead at séances and counted eight million adherents by the end of the century. I don't buy the argument that the lake effect was the catalyst for religious creativity, but it's fun to speculate.

The region also hosted a vibrant antislavery movement. Many abolitionists were women, who craved a more active role than their male colleagues allowed. At the invitation of Elizabeth Cady Stanton, more than two hundred female activists (along with forty men) assembled at the Wesleyan Chapel in Seneca Falls, New York, in July 1848 for the first Woman's Rights Convention. The National Park Service acquired the site in 1985, and the restored chapel is now the centerpiece of Women's Rights National Historical Park. The park's informative exhibits and multimedia program extol the legacy of Stanton and her supporters but say nothing about lake-effect snow.

By contrast, the infamous Seneca Falls winter weather reinforces the area's other, more tenuous claim to fame—as the inspiration for Bedford Falls in Frank Capra's classic film *It's a Wonderful Life*, in which George Bailey, a frustrated small-town banker (played by Jimmy Stewart) survives a suicide attempt thanks to Clarence Odbody, an aptly named underperforming angel who shows George how miserable the community would have been without him. In a triumphant scene toward the end, George rushes into town through a prodigious snowstorm shouting "Merry Christmas!" as he passes a sign reading "Welcome to Bedford Falls." The movie was shot on the RKO backlot in Encino, California, in summer, and the falling snow was an innovative combination of foamite, soap, and water that resembled large, fluffy, gently descending lake-effect snowflakes better than the white-painted corn flakes previously used in the film industry.[3] Seneca Falls boosters also point out that in 1945, a year before filming, director Capra got a haircut in their village on the way to visit an aunt in nearby Auburn. What's more, Seneca Falls is about two hours from Elmira, mentioned perhaps a bit too pointedly by the dour bank examiner, who was eager to wrap up his audit of the Bailey Building & Loan so he could spend the holidays there with his sister.[4] Try to find another small city named Falls within two hours of Elmira.

Oddly, the Seneca Falls Visitors Center, on a website promoting "The Real Bedford Falls," says nothing about snow, aside from noting that the mid-December "It's a Wonderful Life" weekend includes a Snowman Building Contest, and promising that "in the event of lack of snow, substitute materials will be provided."[5] The possibility of no snow might reassure tourists from farther south who recall that, earlier in the film, George crashes his car into a tree during a blinding snowstorm.

Come late January or February, when the snow might have lost its appeal, many of our neighbors depart to the airport for a week or two in Florida, Las Vegas, the Caribbean, or any other place that's warm and sunny. If their return flights are delayed, it's probably not because of snow-covered runways at our local airport, which is better prepared than counterparts in Atlanta and other East Coast hubs, easily stymied by synoptic snowstorms. Those of us who enjoy driving more than flying have other options. When my wife and I can spare the time for a short winter getaway, we check the forecast before heading over to Vermont or New Hampshire in a car with all-wheel drive. Snow is probably on the ground there, and we appreciate New England's picturesque landscapes and cozy inns.

New England might be better known for its photogenic villages, but Upstate New York has its share, with intriguing names like Canandaigua, Cazenovia, and Skaneateles. (Cooperstown could be on the list, but in recent years the tourist traps that glommed onto the Baseball Hall of Fame overran its otherwise pleasant downtown.) And throughout the winter, our "lake-effect machine" (as TV meteorologists insist on calling it) can be relied upon to refresh the snow cover, which otherwise starts looking trashy after a few days. As we've observed more than once in southern Vermont, dirty snow does little for Bennington and Brattleboro.

Fresh and generally persistent snow cover is a boon to upstaters who enjoy snowshoe hiking or cross-country skiing, a less expensive and safer pursuit than downhill skiing, for which snow can be manufactured economically as far south as Maryland and New Jersey as long as temperatures stay a few degrees below freezing. Public trails are a great asset for cross-country enthusiasts and winter hikers, particularly where snowmobiles are excluded.

Don't get me wrong: the snowmobile is a fantastic invention, far more so than its summertime cousin, the jet ski, roundly resented by canoeists, kayakers, and lakeside residents who value solitude. Snowmobiles can be noisy, but they're also fun, as well as a reliable form of local transportation in rural communities when travel is hampered by heavy snow. And they've spawned a significant tourist business in the North Country, particularly in Lewis County, where an extensive network of carefully groomed state-funded trails (fig. 7.5) includes long-ride "corridor" trails that cross county lines. Winter visitors driving large pickups or full-size SUVs bring their machines along in chunky trailers designed to hold two or four snowmobiles. It's a seasonal hobby for folks willing to spend up to $10,000 for a new machine, $2,000 or more for a trailer, and additional amounts for a helmet, warm outerwear, insurance, a registration fee that helps subsidize trail maintenance, and a local trail permit.[6] Snowmobilers from farther south can easily afford overnight accommodations, hearty meals, and membership in an area snowmobile club.[7] Snowmobiling is the largest source of tourism revenue in Lewis County, which also benefits when well-off winter-sports enthusiasts buy second homes, pay property taxes, and don't send kids to local schools.[8]

To collect snapshots for this chapter, my wife and I headed north from Syracuse one dry but cloudy Sunday morning in late February on

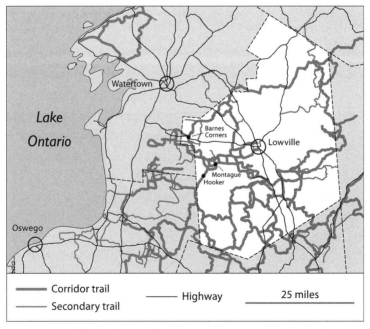

7.5. Snowmobile trails in and around Lewis County, New York, crisscross Tug Hill and converge on Lowville, the county seat.

a 211-mile circuit around Tug Hill through Lowville, and taking in the hamlets of Barnes Corners, Hooker, and Montague, all famous for record snowfalls. Despite several inches of new snow the night before, Interstate 81 and the state highway eastward through Barnes Corners were plowed and dry. At last count, the hamlet held the state record for the largest amount of snow on the ground in any one day: 84 inches on January 25, 1987.[9] Two snowmobiles were parked in the snow next to the defunct gasoline pumps in front of Louie's (fig. 7.6), and several more, along with pickups, SUVs, and trailers, were in the parking lot across the road. No longer in the gasoline business, Louie's is a popular restaurant and convenience store. According to its website, customers can buy "trail maps, oil, spark plugs, and headlight and taillight bulbs for most snowmobiles" as well as propane, a cleaner, less expensive fuel used by some snowmobiles.[10]

Our next stop was Hooker, which holds the official state record for the greatest monthly snowfall (182 inches in January 1978) as well as the

7.6. Louie's restaurant and convenience store, Barnes Corners, New York, in western central Lewis County.

greatest seasonal snowfall (379.5 inches between August 1977 and July 1978). Four miles east of Barnes Corner we turned off the state highway onto a local road with hard-packed snow—not ice, which would have been treacherous—and drove south. Snowmobiles headed in the opposite direction were four times as numerous as cars: on some rural roads a widely plowed right-of-way doubles as an ad hoc snowmobile trail. A half-hour later we reached what was apparently Hooker: a few structures, presumably houses, but many of them vacant, and no revealing labels on buildings or road signs. I checked my maps and confirmed that this was indeed Hooker, and the GPS agreed. Although a collage of signs (fig. 7.7) warned that the route ahead was not only a dead end but also a "minimum maintenance road," a small brown-and-yellow sign resembling the Interstate Highway shield beckoned snowmobilers to continue south on Corridor 5.

Hooker is definitely at the lower end of the hamlet spectrum, and certainly not a village. The US Geological Survey's Geographic Names Information System generously calls it a "populated place" and provides

7.7. Signs mark the seasonal end of the road at Hooker, New York.

a precise latitude and longitude but little more.[11] The hamlet might have been named for a nineteenth-century county resident—John Hooker or Urial Hooker—not the lax and ineffective Union Army general Joseph Hooker, whose encampments were famous for attracting prostitutes.[12] (Googling "Hooker, New York" turned up websites focused on escort services or former New York governor Eliot Spitzer, who resigned in March 2008, after only fourteen months in office, because of a sex scandal.) Co-op snowfall observations have been taken at or near Hooker since 1912. NCDC data indicate that the record snowfall was actually measured at "Hooker 12 NNW," twelve miles north of the hamlet, on the fringe of Lookout State Forest.[13]

Our next stop was Montague, site of a NEXRAD radar tower and famous for the disputed and apparently erroneous record 77-inch one-day snowfall reported in 1997. We had to backtrack four miles, which afforded an opportunity to photograph a highway department plow shoving snow aside while making little attempt to expose the pavement (fig. 7.8) and a snowmobiler speeding along on the adjacent trail (fig. 7.9). At Montague twenty snowmobiles were parked outside the Montague Inn (fig. 7.10), and

7.8. Snowplow at work north of Hooker.

7.9. Snowmobile on the C5 snowmobile trail adjacent to the town road, north of Hooker, New York.

7.10. Snowmobiles parked at the Montague Inn.

several SUVs with trailers attached were parked next to the motel annex, across the road. The inn's website notes that it is "located on Snowmobile Trails C8A and S55"—C for corridor, S for secondary.[14] Signs identify other trails through the intersection and remind riders that trail groomers—big machines on crawler treads—have the right-of-way (fig. 7.11).

The inn was thriving that day, but when we passed by at roughly the same time one Sunday the previous August, there were only two vehicles in the parking lot. The property has been on the market for several years, and a small "For Sale" sign on the front of the building signaled the present owner's eagerness to move on. To help largely seasonal businesses like the inn, Lewis County officials have been promoting various forms

7.11. Trail signs in front of the Montague Inn include an insurance company's "Emergency Locator" with GPS coordinates.

of non-winter recreation including ATV (all-terrain vehicle) riding, with its own trail system. Although ATVs can be fitted with tracks for use on groomed trails, they're not welcome on state-supported snowmobile trails.

Leaving Montague, we drove north to the well-plowed state highway, and then headed east toward Lowville, past the vast Maple Ridge Wind Farm, where 195 wind turbines on 260-foot towers take advantage of Tug Hill's geography and augment the local tax base.[15] I couldn't resist photographing a partly collapsed farmhouse near the road (fig. 7.12). Whether its decline began when heavy snow collapsed the roof is difficult to discern, but the snow's weight must partly account for its present state. Andrew Wyeth would have appreciated it.

Lewis County's windmills might suggest prosperity, but the county seat's distressed downtown demonstrates the typical result when a strip mall with abundant parking pops up near the edge of an otherwise stable village. The Federal Writers Project guide to New York, prepared in the late 1930s, describes Lowville as "a characteristic North Country village with tree-shaded streets, frame dwellings, and a neat brick business section,"[16] but vacant storefronts now exacerbate what locals call the "gap in

7.12. Collapsed house buried in snow west of Lowville.

the smile of Lowville" left by a 1999 fire that destroyed a block of five buildings.[17] Out-of-town recreationalists probably couldn't care less—the "Welcome, Snowmobilers" sign (fig. 7.13) at Stewart's, a chain convenience store between downtown and the strip mall promises a friendly human smile as well as cheap gas and decent take-out coffee. Snowmobilers don't savor New England quaint.

A large sign near the village line contradicts the stigma of a waning commercial core (fig. 7.14). Beneath the banner "Welcome to Lowville, Est. 1847" an array of fifteen placards touts a vibrant community with churches for seven different denominations (Baptist, Episcopal, Methodist, Nazarene, Presbyterian, Roman Catholic, and the "New Day Community Church"), three fraternal organizations (Elks, Lions, and Masons), three veterans groups (American Legion, Marine Corps League, and VFW), the Lowville Business Association, and Operation Lowville, a now-defunct nonprofit organization that had sponsored concerts, festivals, and other arts events.[18]

7.13. A local snowmobile club and a Lowville convenience store appreciate snow tourists.

7.14. A welcome sign advertises an array of community organizations.

Lowville can no longer claim what the Writers' Project called "one of the largest cold-storage plants in the world," but Kraft Foods, which bought the company, still makes cream cheese here at an updated facility and is a major local employer and buyer of raw milk.[19] Although Lowville now hosts the only remaining US plant making wooden bowling pins— thanks to consolidation and outsourcing—chronic unemployment in manufacturing underscores the local importance of snow tourism.[20]

Communities throughout the Great Lakes snowbelts share Lowville's rust-belt woes, which are largely a consequence of economic restructuring rather than aversion to snow. Although many retirees head south for a warmer climate, summer in the South can be far more repugnant than winter in the North, and those who can afford a condo in south Florida often become snowbirds, fleeing in December and returning in April, but not always in time for the decennial census enumeration. While the US population grew by 9.7 percent between 2000 and 2010, most of the country's 3,143 counties either registered a net loss or grew more slowly than the nation as a whole.[21] If there's even a weak correlation with lake-effect snow, it's not apparent on the map of population change in the Great Lakes region (fig. 7.15) that I compiled by extracting boundaries and categories from a Census Bureau graphic, generalizing its color symbols to graytones and adding dotted lines encompassing Val Eichenlaub's snowbelts (fig. 1.1) to show where the lake effect is most pronounced. Only two counties within a dotted line are shown in solid black, indicating an above-average increase—both are on the western side of Michigan's lower peninsula—and only two other above-average counties border the Lakes. Even so, slow-growth and no-growth counties abound throughout the Northeast and Midwest, while high-growth counties are common in the outer suburbs of thriving metropolitan areas like Chicago, Minneapolis, and Philadelphia.

Slow growth is not necessarily bad, according to *Bloomberg Businessweek*, which included Marquette County, Michigan, and Onondaga County, New York, in its list of America's "best affordable places." Though neither is a retirement utopia or a metropolitan magnet, these

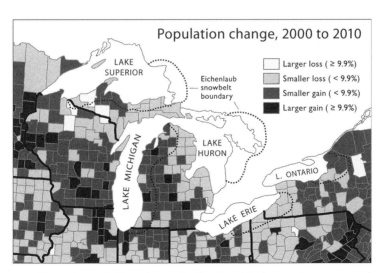

7.15. Recent population change in the Great Lakes region compared to Val Eichenlaub's snowbelts.

snowbelt counties offer jobs and affordable housing to the young profes-
sionals that social scientist Richard Florida labeled the creative class.[22]
Vignettes on the *Businessweek* website highlight below-average rents and
home prices, below-average unemployment, and above-average propor-
tions of adult residents with a bachelor's degree—at 28.6 and 31.5 percent,
respectively, Marquette and Onondaga counties are comfortably above
the national rate of 27.5 percent, and well above Lewis County's 14.3 percent.[23] Major employers like Marquette General Health System, Cliffs
Natural Resources (CLF), Michigan Operations, and Northern Michigan
University in the Upper Peninsula, and Upstate University Health System
and Syracuse University in Central New York underscore the importance
of health care, higher education, and research. Steady expansion in these
sectors has at least partly compensated for industrial decline without over-
heating the local real-estate market.

In Florida's narrative, the creative class values outdoor recreation, a
good environment for raising children, and a cultural mix that includes
the arts, a music scene, and innovative restaurants. When balanced by
vibrant urban enclaves like Downtown Marquette and Syracuse's Armory
Square, lake-effect snow can make in-your-face seasonality an asset. That's
a rallying cry of "40 Below," a support group of over 600 young adults, 97
percent with college degrees, formed in Syracuse in 2004 with support
from the Metropolitan Development Association.[24] Several of their 'Cuse
Comebacks, compiled to counter a range of slurs, debunk assertions that
"the weather sucks."[25]

> I think we get the most snow out of all Central Upstate and maybe that's
> what we should be proud of. We should say, "Yeah, we get the most,
> good for us."
> —NICHOLE WENDERLICH, 26, chair of the 40 Below
>   marketing and communications task force

> We have no natural disasters. We never have to worry about hurricanes.
> Very rarely a flood. I can live with the occasional snow storm. Then you
> can get out there and ski.
> —ABBEY DALY, 29, real estate broker at Coldwell Banker

The weather here is beautiful. The four seasons make you stop and think about what nature has provided us. We have access to such a variety of activities because of the change of seasons.
>   —CHUCKIE HOLSTEIN, 81, executive director of FOCUS
>   Greater Syracuse

It can be a challenge to live here in the dead of winter, but I do like the change of seasons that we experience in Upstate New York. Every season has its own color and beauty . . . and of course, clothing.
>   —THERESA BARRY, 42, owner of Razz Barry boutique in
>   Armory Square

We have just enough variety in the weather to keep things interesting. Sure, there's never enough of that one season you love, but that makes you more appreciative of the good days.
>   —FRANK CALIVA, 48, director of Talent Initiatives for
>   Metropolitan Development Association

Granted, there are times the weather is bad, but I don't think people take the time to notice the nice weather. Everyone just associates this area with the snow.
>   —DOUG STRAHLER, 24, partner of Threeonefive Design

I think that in Central New York and Upstate New York there is nowhere more beautiful in September, October and November.
>   —BETH SAVICKI, 32, marketing communications specialist for
>   the Greater Syracuse Chamber of Commerce

Recent efforts to celebrate rather than apologize for our winter weather include Upstate Snowdown, an early February 2011 fund-raiser in which teams competed in a snow sculpture contest, members sold hot chocolate and cupcakes, and children lobbed snowballs at volunteers dressed as a cow and a lobster.[26] Later that month the eleven-day Winterfest, sponsored by numerous businesses and public organizations, invited people downtown to try out their ice skates, see the sixteen-foot ice castle and other exhibits, and sample entries in the annual chili cook-off and chowder competition.[27] As often happens, Mother Nature intervened with a short but inappropriate warm spell.

Anne Mosher, a cultural-historical geographer who's a colleague at SU, has studied the area's seasonality in several contexts, past and present. For her current project, a critical look at the Erie Canalway National Heritage Corridor, a federal effort to promote economic development along the historic waterway, she found a strong "April to November bias" that represents the region as "always sunny and summer."[28] In all of their place-promotion campaigns, state and federal governments largely ignore our pervasive seasonality, except to highlight skiing or snowmobiling, while the national news media, both print and electronic, promote "the image of Upstate as buried in snow." By contrast, local TV meteorologists relish the opportunity to brandish their knowledge of lake-effect storms and the prowess of their visualization software and in-house radar systems; what Anne calls their "swagger" has made local viewers intimately, if not always knowledgeably, aware of lake snow as a force in their lives. The result is two very different "place stereotypes," one for residents, the other for outsiders.

Although the lake effect makes regular appearances on local TV news in the Great Lakes states as early as October or as late as May, its cameo appearances before a national audience, from December through February, have a wider impact. Newspapers outside the region and the national networks' nightly news programs notice only extreme weather events like the twister that cuts a corridor across a county or the hurricane that topples beachfront homes into the sea; in an era of terrorist bombings and financial and nuclear meltdowns, seventeen consecutive days of snow flurries is hardly newsworthy. Sadly, the lake effect's prime opportunity for nationwide notoriety is the rare whiteout that causes a fifty-car pileup or the unusual multiday storm that breaks a record or immobilizes a major city—events that warrant from-the-scene reporting by the Weather Channel, CNN, and *Good Morning America*. Although committed TWC viewers are no doubt aware that lake-effect snow is neither exceptional nor inherently brutal, the rest of the country too easily lumps it with the disastrous weather of Tornado Alley and the Gulf Coast. "It snows there all the time," outsiders know. "How can people stand it?"

What the national media also miss is the lake effect's gentler side during what Val Eichenlaub called the "stable season."[29] From mid-March through August, when Lake Michigan is cooler than the surrounding land, relatively cool winds off the lake bring comparatively little moisture and more moderate summer temperatures, thereby reducing the frequency of summer thunderstorms. Farther north, around Lake Superior, the stable season arrives later, while around Lake Erie, which warms more readily because it is smaller and shallower, the effect arrives and departs earlier. And because the lakes also lessen the likelihood of late frosts in spring and early frosts in fall, apple orchards abound east of Lake Michigan as well as southeast of Lake Erie in Pennsylvania and New York and along the southern shore of Lake Ontario from Hamilton, Ontario, to Oswego, where the snowbelt doubles as a fruit belt. Overall an agreeable summer and a pleasant fall compensate nicely for an obstinate winter.[30]

Notes

Illustration Credits

Index

Notes

1. Recipe

1. I'm not making this up; see Carl F. Ojala, "A Case for Multi-Scale Analysis: The 'Men's Pants Belt' of Northern Georgia," *Geographical Survey* 5, no. 3 (1976): 17–24.

2. Val L. Eichenlaub, "Lake Effect Snowfall to the Lee of the Great Lakes: Its Role in Michigan," *Bulletin of the American Meteorological Society* 51 (1970): 403–12; map on 404.

3. I obtained averages for the thirty-year period bracketed by the 1930–31 and 1959–60 snow seasons from the Northeast Regional Climate Center at Cornell University. Though useful, early and average amounts like these can ignore marked variations throughout an urban area as well as over time, especially when the measurement site is relocated from downtown to the airport.

4. Val L. Eichenlaub, *Weather and Climate of the Great Lakes Region* (Notre Dame, IN: Univ. of Notre Dame Press, 1979); map on 165.

5. Great Lakes of North America Snow Belt, http://en.wikipedia.org/wiki/File:Great _Lakes_Snowbelt_EPA_fr.png. It's NOAA's National Weather Service, not that EPA, that collects US snowfall data. Perhaps Pierre confused the EPA with the EDS (Environmental Data Service), which was consolidated with the NWS and other offices when NOAA was formed in 1970.

6. Eichenlaub, *Weather and Climate of the Great Lakes Region*, 153–55.

7. For the 250 largest US cities the National Climatic Data Center lists average total snowfall for periods of variable length ending in 2002. http://lwf.ncdc.noaa.gov/oa /climate/online/ccd/snowfall.html. I rounded the NCDC amounts to the nearest whole inch. Toronto's average, for 1971–2000, is from Environment Canada, http://www.climate .weatheroffice.gc.ca/climate_normals/stnselect_e.html.

8. In snowbelt areas, 30 to 60 percent of annual snowfall can be attributed to lake-effect storms. William Agrella, "Lake Effect Snowfall in Western New York and Surface Temperatures of Lakes Erie and Ontario" (master's thesis, Texas A&M Univ., 1978), 107.

9. Greg E. Mann, Richard B. Wagenmaker, and Peter J. Sousounis, "The Influence of Multiple Lake Interactions upon Lake-Effect Storms," *Monthly Weather Review* 130 (2002): 1510–30.

10. Depth figures are from the Great Lakes Information Network website, http://www.great-lakes.net/lakes/ref/lakefact.html.

11. Raymond Assel, Kevin Cronk, and David Norton, "Recent Trends in Laurentian Great Lakes Ice Cover," *Climatic Change* 57 (2003): 185–204. Ice-cover percentages are medians for the period 1962–2001. Mean maximum ice cover for Lakes Erie and Ontario were 88 and 28 percent, respectively.

12. Jason M. Cordeira and Neil F. Laird, "The Influence of Ice Cover on Two Lake-Effect Snow Events over Lake Erie," *Monthly Weather Review* 139 (2008): 2747–63. The effect of ice cover seems noteworthy only when coverage exceeds 70 percent; see Mathieu R. Gerbush, David A. R. Kristovich, and Neil F. Laird, "Mesoscale Boundary Layer and Heat Flux Variations over Pack Ice-Covered Lake Erie." *Journal of Applied Meteorology and Climatology* 47 (2008): 668–82.

13. Data are from the Buffalo office of the National Weather Service, http://www.erh.noaa.gov/buf/climate/buf_snownorm.htm.

14. For a description of the development and impact, see National Weather Service, Buffalo, "Lake Effect Storm 'Chestnut,' November 20–23, 2000," http://www.erh.noaa.gov/er/buf/lakeffect/lakeo001/c/stormcsum.html.

15. For a concise discussion of the principles and history of weather radar, see Mark Monmonier, *Air Apparent: How Meteorologists Learned to Map, Predict, and Dramatize Weather* (Chicago: Univ. of Chicago Press, 1999), esp. 137–52. An intensive training course helps forecasters appreciate the complexity of radar images, which do not show what is falling from the cloud or reaching the ground. Although weather radars are designed to assume spherical targets of liquid water, a trained forecaster must judge the impacts of tiny ice crystals, rather than liquid water, on the map of reflectivity.

16. Robert M. Rauber, John E. Walsh, and Donna J. Charleviox, *Severe and Hazardous Weather: An Introduction to High Impact Meteorology*, 2nd ed. (Dubuque, IA: Kendall/Hunt, 2005), 209–11.

17. Although snowbands mimic the convective structure of thunderstorms, lake-effect snowstorms are somewhat similar to hurricanes, which are much larger systems and circular rather than elongated. Both systems typically form over large bodies of warm water, which provide the heat and moisture needed to survive. Hurricanes can travel great distances across the Atlantic Ocean, but over land they weaken and eventually die. Similarly, lake storms also weaken with increased distance from the lake.

18. Douglas B. Carter, "Climate," in *Geography of New York State*, ed. John H. Thompson, 54–78 (Syracuse: Syracuse Univ. Press, 1966), esp. 69 and fig. 22 (folded map in pocket).

19. For examples of Tug Hill residents' perverse pride in their challenging winters, see Matt J. Macierowski, *Lake Effect Snows East of Lake Ontario* (Boonville, NY: Boonville Graphics, 1979); and Harold E. Samson, *Tug Hill Country: Tales for the Big Woods* (Lakemont, NY: North Country Books, 1971).

20. Bob Swanson and Adrienne Lewis, "Many Ingredients Go into Lake-Effect Snow," *USA Today*, November 5, 2008; and *Encyclopedia of Earth*, s.v. "Lake Effect Snow," http://www.eoearth.org/article/Lake_effect_snow.

21. Eichenlaub, *Weather and Climate of the Great Lakes Region*, 153–55.

22. Laurence W. Sheridan, "The Influence of Lake Erie on Local Snows in Western New York," *Bulletin of the American Meteorological Society* 22 (1941): 393–95.

23. B. L. Wiggin, "Great Snows of the Great Lakes," *Weatherwise* 3 (1950): 123–26.

24. Thomas A. Niziol, "Operational Forecasting of Lake Effect Snowfall in Western and Central New York," *Weather and Forecasting* 2 (1987): 310–21; diagram on 315.

25. I simplified Niziol's comparatively precise condition by substituting compass directions for his azimuths, measured in degrees clockwise from north, and ranging from $230°$ to $340°$ on Lake Erie and from $230°$ to $80°$ on Lake Ontario.

26. Meteorologist Jeff Haby, "Lake Effect Snow Questions," http://www.theweather prediction.com/winterwx/lesnow/tree/.

27. Eichenlaub, "Lake Effect Snowfall to the Lee of the Great Lakes," 404.

28. Tayfun Kindap, "A Severe Sea-Effect Snow Episode over the City of Istanbul," *Natural Hazards* 54 (2010): 707–23.

29. Jeff S. Waldstreicher, "A Foot of Snow from a 3000-foot Cloud: the Ocean-Effect Snowstorm of 14 January 1999," *Bulletin of the American Meteorological Society* 83 (2002): 19–22.

30. Neil Laird, Ryan Sobash, and Natasha Hodas, "The Frequency and Characteristics of Lake-Effect Precipitation Events Associated with the New York State Finger Lakes," *Journal of Applied Meteorology and Climatology* 48 (2009): 873–84.

31. Melissa Payer, Jared Desrochers, and Neil F. Laird, "A Lake-Effect Snowband over Lake Champlain," *Monthly Weather Review* 135 (2007): 3895–3900.

2. Discovery

1. J. H. Mather and L. P. Brockett, *Geography of the State of New York, Embracing Its Physical Features, Climate, Geology, Mineralogy, Botany, Zoology, History, Pursuits of the People, Government, Education, Internal Improvements, &c.* (Hartford, CT: J. H. Mather and Co., 1847).

2. J. H. French, *Gazetteer of the State of New York: Embracing a Comprehensive View of the Geography, Geology, and General History of the State, and a Complete History and Description of Every County, City, Town, Village, and Locality* (Syracuse: R. Pearsall Smith, 1860), esp. 20–23.

3. E. T. Turner, *The Climate of the State of New York*, in *Fifth Annual Report of the Meteorological Bureau and Weather Service of the State of New York*, 1893 (Albany: James B. Lyon, State Printer, 1894), 345–457.

4. E. T. Turner, "The Physical Geography of New York State. Part XI. The Climate of New York," *Journal of the American Geographical Society of New York* 32 (1900): 101–32; and his "The Climate of New York," in Ralph S. Tarr, *The Physical Geography of New York State*, 331–66 (New York: Macmillan, 1902).

5. Christopher A. Fiebrich, "History of Surface Weather Observations in the United States," *Earth Science Reviews* 93 (2009): 77–84.

6. Everett Mendelsohn, "John Lining and His Contribution to Early American Science," *Isis* 51 (1960): 278–92.

7. William C. Redfield, "Remarks on the Prevailing Storms of the Atlantic Coast, of the Northeastern States," *American Journal of Science* 20 (1831): 11–51.

8. William C. Redfield, "Observations on the Storm of December 15, 1839," *Transactions of the American Philosophical Society* 8 (1843): 77–80. For a history of the controversy, see James Rodger Fleming, *Meteorology in America, 1800–1870* (Baltimore: The Johns Hopkins Univ. Press, 1990), esp. 23–53.

9. Eric R. Miller, "The Evolution of Meteorological Institutions in the United States," *Monthly Weather Review* 59 (1931): 1–6.

10. "Third Report of the Committee on Meteorology," *Journal of the Franklin Institute* 19 (1837): 17–21; and James P. Espy, *The Philosophy of Storms* (New York: Charles C. Little and James Brown, 1841), 105. For discussion of the map's significance and originality, see Fleming, *Meteorology in America*, 58–59; and Armand N. Spitz, "Meteorology at the Franklin Institute," *Journal of the Franklin Institute* 237 (1944): 271–87, 331–57, esp. 280–82.

11. Edgar Erskine Hume, "The Foundation of American Meteorology by the United States Army Medical Department," *Bulletin of the History of Medicine* 8 (1940): 202–38.

12. Franklin R. Hough, *Results of a Series of Meteorological Observations Made in Obedience to Instructions from the Regents of the University at Sundry Academies in the State of New York, from 1826 to 1850, Inclusive* (Albany: Weed, Parsons and Company, 1855).

13. For discussion of resultant (prevailing) wind directions, see Hough, *Results of a Series of Meteorological Observations*, xi, 502.

14. Ibid., 463.

15. For time-series graphs summarizing the evolution of snowfall and snow-day observations in the United States, see Charles Franklin Brooks, "The Snowfall of the Eastern United States," *Monthly Weather Review* 43 (1915): 3–11.

16. US Surgeon-General, *Meteorological Register for the Years 1822, 1823, 1824 & 1825, from Observations Made by the Surgeons of the Army, at the Military Posts of the United States* (Washington, DC: Edward De Krafft, 1826), summary table on 63. Unlike later statistics, the yearly averages here are reported for an average month, rather than for an average year, and thus must be multiplied by 12.

17. US Surgeon-General's Office, *Meteorological Register, for Twelve Years, from 1843 to 1854, Inclusive, from Observations Made by the Officers of the Medical Department of*

the Army, at the Military Posts of the United States (Washington, DC: A. O. P. Nicholson, 1855), quotations on vi and viii.

18. Ibid., 666. The range is based on yearly summaries; the report did not include an average annual number of snow days. There were no records for Fort Brady for 1849 and the last seven months of 1848. It had the highest tally for most years, including 1854, but other military posts that reported for only part of the period occasionally registered a greater number of snow days. In 1843, for instance, Buffalo Barracks (missing after 1844) had 67 snow days, notably more than Fort Brady's 43.

19. Ibid., 647.

20. Ibid., 737.

21. Martin A. Baxter, Charles E. Graves, and James T. Moore, "A Climatology of Snow-to-Liquid Ratio for the Contiguous United States," *Weather and Forecasting* 20 (2005): 729–44, esp. map on 732. Also see A. Boyd Pack, "The Water Content of Snow-storms in New York State: Variations among Different Physiographic Regions," *Proceedings of the 26th Eastern Snow Conference, Portland, Maine, 6–7 Feb 1969*, 46–54.

22. US Surgeon-General's Office, *Meteorological Register, for Twelve Years, from 1843 to 1854*, 759.

23. Ibid., 750.

24. Ibid., 753.

25. Ibid., iv.

26. Lorin Blodget, *Climatology of the United States, and of the Temperate Latitudes of the North American Continent, Embracing a Full Comparison of These with the Climatology of the Temperate Latitudes of Europe and Asia, and Especially in Regard to Agriculture, Sanitary Investigations, and Engineering, with Isothermal and Rain Charts for Each Season, the Extreme Months, and the Year* (Philadelphia: J. B. Lippincott and Co., 1857).

27. Fleming, *Meteorology in America*, 110–11; and Marc Rothenberg et al., eds., *The Papers of Joseph Henry*, vol. 8, *The Smithsonian Years, January 1850–December 1853* (Washington: Smithsonian Institution Press, 1998), xxii and 274.

28. For fuller discussion of the dispute between Henry and Blodget, see Fleming, *Meteorology in America, 1800–1870*, 111–15; and Marc Rothenberg et al., eds., *The Papers of Joseph Henry*, vol. 9, *The Smithsonian Years, January 1854–December 1857* (Washington, DC: Smithsonian Institution Press, 2002), xxvi and 197–204. For Blodget's acknowledgments of the Smithsonian Institution, see his *Climatology*, vi (twice), vii, viii, 33, 37, and 233.

29. US Surgeon-General's Office, *Meteorological Register, for Twelve Years, from 1843 to 1854*, 736; Blodget, *Climatology of the United States*, 319.

30. Blodget, *Climatology of the United States*, 337, 340, 348–49.

31. Ibid., 345.

32. Anonymous, "Climatology of the United States . . . [review]," *North American Review* 91:327–54, quotations on 348–49. Dewey's authorship of the anonymous review is revealed in the cumulative index to volumes 1–125 of the *Review*, published in 1878.

33. Charles A. Schott, *Tables and Results of the Precipitation, in Rain and Snow, in the United States: and at Some Stations in Adjacent Parts of North America, and in Central and South America*, Smithsonian Contributions to Knowledge 222 (Washington, DC: Smithsonian Institution, 1872), 120.

34. Ibid., 121.

35. Copper Falls Mines and Eagle River Mine, relatively remote sites now considered ghost towns, registered winter averages of 13.44 and 10.85 inches, respectively; ibid., 56–59.

36. New York Meteorological Bureau and Weather Service, *Fifth Annual Report, 1893* (Albany: James B. Lyon, 1894), n.p. (back of volume).

37. Table 24 in the New York report lists monthly, seasonal, and annual precipitation averages as well as the period covered for 80 stations; ibid., 414–17. A similar table in the Smithsonian report identifies 164 stations, including 32 without winter averages. Schott, *Tables and Results of the Precipitation, in Rain and Snow*, 22–30.

38. Brooks, "Snowfall of the Eastern United States," 2–3. Brooks differs by a year from a Weather Bureau report that noted "no measurements were made of the depth of unmelted snow . . . until the early part of 1883." US Weather Bureau, *Report of the Chief of the Weather Bureau, 1891–92* (Washington, DC, 1892), 447.

39. W. B. Hazen, *History of the Signal Service with Catalogue of Publications, Instruments and Stations* (Washington, DC: US Signal Service, 1884), 38; and Miller, "Evolution of Meteorological Institutions in the United States." Also see W. R. Baron, "Retrieving American Climate History: A Bibliographic Essay," *Agricultural History* 63 (1989): 7–35.

40. US Signal Service, *Annual Report of the Chief Signal Officer of the Army to the Secretary of War for the Year 1886* (Washington, DC: Government Printing Office, 1886), 291.

41. Mark W. Harrington, *Rainfall and Snow of the United States, Compiled to the End of 1891, with Annual, Seasonal, Monthly, and Other Charts*, Bulletin C (Washington, DC: Weather Bureau, 1894).

42. Ibid., 16.

43. Ibid.

44. Frank Waldo, *Elementary Meteorology for High Schools and Colleges* (New York: American Book Company, 1896), 345.

45. A. J. Henry, "Normal Annual Sunshine and Snowfall," *Monthly Weather Review* 26 (1898): 108. Henry had discussed his snowfall chart briefly a month before in his "Meteorological Tables and Charts," *Monthly Weather Review* 26 (1898): 68–69.

46. Even earlier, the *Monthly Weather Review* had included maps of snow on the ground at the end of the month.

47. For a list of climate stations and a map of their locations, see Alfred Judson Henry, *Climatology of the United States*, Weather Bureau Bulletin Q (Washington, DC: Government Printing Office, 1906), 113–18 and plate xxxiii.

48. Ibid., 58–59. Details in Henry's discussion indicate that he did not merely summarize his 1898 map.

49. Charles F. Brooks, "The Snowfall of the United States," *Quarterly Journal of the Royal Meteorological Society* 39 (1913): 81–86.

50. Charles Franklin Brooks, "The Snowfall of the Eastern United States" (PhD diss., Harvard Univ., 1914), 18.

51. Ibid.

52. Ibid.

53. Ibid., 19.

54. Ibid., 30–31.

55. Ibid., 32.

56. Charles Franklin Brooks, "The Distribution of Snowfall in Cyclones of the Eastern United States," *Monthly Weather Review* 42 (1914): 318–30; "The Snowfall of the Eastern United States," *Monthly Weather Review* 43 (1915): 2–11, plus 15 charts; and "New England Snowfall," *Geographical Review* 3 (1917): 222–40.

57. For an online collection of Brooks's columns, see National Oceanic and Atmospheric Administration, NOAA Central Library, Digital Documents and Maps Collection, Why the Weather? http://docs.lib.noaa.gov/rescue/whytheweather/whytheweather .html. The column was continued beyond 1927 by several Weather Bureau personnel. Brooks's installments were issued as a book; see his *Why the Weather?* (New York: Harcourt, Brace and Co., 1924, rev. 1935).

58. Charles F. Brooks and A. J. Connor, *Climatic Maps of North America* (Cambridge, MA: Blue Hill Meteorological Observatory of Harvard Univ. and Harvard Univ. Press, 1936), quotation from inside front cover.

59. Among the volumes completed was Robert DeC. Ward, Charles F. Brooks, and A. J. Connor, *The Climates of North America*, vol. 2, part J of *Handbuch der Klimatologie*, ed. Wladimir Köppen and Rudolph Geiger (Berlin: Verlag von Gebrüder Borntraeger, 1936).

60. Everett E. Edwards, *A Bibliography of the History of Agriculture in the United States* (New York: Burt Franklin, 1970), 19; and S. S. Visher and Charles Y. Hu, "Oliver Edwin Baker, 1883–1949," *Annals of the Association of American Geographers* 40 (1950): 328–34.

61. Robert DeC. Ward, "The New Precipitation Section of the Atlas of American Agriculture," *Monthly Weather Review* 49 (1922): 117–24; quotation on 117.

62. J. B. Kincer, Advance sheets 5, part 2, Climate; section A, Precipitation and Humidity, *Atlas of American Agriculture* (Washington, DC: Government Printing Office, 1922), fig. 84 on 44.

63. Kincer, *Atlas of American Agriculture*, 43–44.

64. J. B. Kincer, "Climate and Weather Data for the United States," in US Department of Agriculture, *Climate and Man*, 1941 Yearbook of Agriculture, 685–747 (Washington, DC: Government Printing Office, 1941).

65. Stephen Sargent Visher, *Climatic Atlas of the United States* (Cambridge, MA: Harvard Univ. Press, 1954), quotations on 17. Although it includes some original drawings, the *Atlas* is mostly a compilation of climate maps from various reliable sources. Visher redrew many of Kincer's maps, most noticeably the snowfall map from *Climate and Man* for map 592 in his *Atlas* (on page 232) and the snow-days maps from the *Atlas of American Agriculture* for map 596 (on page 233).

66. US Geological Survey, *National Atlas of the United States* (Washington, DC, 1970), 100.

67. U.S. Environmental Data Service, *Climatic Atlas of the United States* (Washington, DC: Government Printing Office, 1968), 53. In a round of optimistic reorganization, the Congress had formed ESSA by combining the Weather Bureau (moved from the Department of Agriculture to the Department of Commerce in 1940) with the US Coast and Geodetic Survey. ESSA was renamed NOAA (National Oceanic and Atmospheric Administration) in 1970. EDS was renamed the Environmental Data and Information Service in 1978 and eventually became part of the National Environmental Satellite, Data, and Information Service (NESDIS), which includes the National Climatic Data Center, formed in Asheville, North Carolina, in 1951, to consolidate weather records from the Weather Bureau and the military.

68. There's also a 256-inch snowfall line, used only in the West.

69. The strategy shifted to multiples of 100 for the mountainous, rugged West, where 200-, 300-, and 400-inch lines appear.

70. John H. Thompson, ed., *Geography of New York State* (Syracuse: Syracuse Univ. Press, 1966).

71. Robert A. Muller, "Snowbelts of the Great Lakes," *Weatherwise* 19 (1966): 248–55; quotation on 252. I also interviewed Muller, by telephone, on April 26, 2010. Muller's two-page map is noticeably more detailed in the East, from Pennsylvania to Maine, than in the Midwest, from Ohio to Minnesota.

72. Eichenlaub, "Lake Effect Snowfall to the Lee of the Great Lakes," 403–12; map on 404.

73. For examples, see the US Geological Survey's National Seismic Hazard Maps website, http://earthquake.usgs.gov/hazards/products/.

74. Richard P. Cember and Daniel S. Wilks, *Climatological Atlas of Snowfall and Snow Depth for the Northeastern United States and Southeastern Canada*, Research Series publication no. RR 93-1 (Ithaca, NY: Northeast Regional Climate Center, 1993).

75. Robert W. Scott and Floyd A. Huff, "Impacts of the Great Lakes on Regional Climate Conditions," *Journal of Great Lakes Research* 22 (1996): 845–63.

3. Prediction

1. "Mitchell Retires at Central Office," *Weather Bureau Topics*, July 1950, 104–5. Also see Gordon Dunn and R. Cecil Gentry, "Forecaster's Biography: Charles L. Mitchell: Remarkable Forecaster—Rare Friend," *Weather and Forecasting* 1 (1986): 108–10.

2. "Retirements: Wilfred P. Day," *Weather Bureau Topics*, April 1957, 78–79. Also see "Weatherman 40 Years, He Calls It a (Nice) Day," *Washington Post*, March 1, 1957, A3.

3. C. L. Mitchell, "Snow Flurries along the Eastern Shore of Lake Michigan," *Monthly Weather Review* 49 (1921): 502–3.

4. US Bureau of the Census, *Official Register of the United States*, 1921 (Washington, DC: Government Printing Office, 1922), 372, 871.

5. Mitchell, "Snow Flurries along the Eastern Shore of Lake Michigan."

6. R. M. Dole, "Snow Squalls of the Lake Region," *Monthly Weather Review* 56 (1928): 512–13.

7. Mark Monmonier, *Air Apparent: How Meteorologists Learned to Map, Predict, and Dramatize Weather* (Chicago: Univ. of Chicago Press, 1999), esp. 1–17, 57–80.

8. Sheridan, Laurence W., "The Influence of Lake Erie on Local Snows in Western New York," *Bulletin of the American Meteorological Society* 22 (1941): 393–95.

9. Details of Sheridan's work in meteorology are sketchy. I relied largely on his listing in the 12th edition of *American Men and Women of Science* (1971–73); his obituary in the October 1, 1984, issue (p. B3) of the *Altoona Mirror*; and his daughter, Christine Roth, who shared recollections of her father's career. The catalog of the NOAA Central Library lists him as author of an unpublished Weather Bureau study on "The use of ocean currents in long range forecasting" (undated) and co-author (with Harry Wexler) of "A study of barograph records for July 16, 1945, in the vicinity of the New Mexico atomic bomb explosion" (1946). Though not listed in the catalog, he also published "A Study of Northern Hemisphere Pressure-Center Tracks," *Transactions of the American Geophysical Union* 26 (1): 49–57. After leaving the Weather Bureau, Sheridan taught mathematics at St. Thomas College, in St. Paul, Minnesota. From 1952 to 1960, he worked in industry, as a physicist and engineer, before joining the mathematics faculty at the Pennsylvania State University's Altoona campus in 1960.

10. John T. Remick, "The Effect of Lake Erie on the Local Distribution of Precipitation in Winter (I)," *Bulletin of the American Meteorological Society* 23 (1942): 1–4.

A month later a second short installment—same title followed by (II)—addressed the impact of two severe December storms, in 1930 and 1937.

11. For details of John Remick's life and career, I relied principally on the World War II Army Enlistment Records, online at the Access to Archival Databases (AAD) website of the National Archives, *aad.archives.gov*; his obituary in the April 1961 issue (p. 69) of *Weather Bureau Topics*; and his nephew, James K. Remick, of Lockport.

12. Victor P. Starr, *Basic Principles of Weather Forecasting* (New York: Harper and Brothers, 1942), esp. 49–51; citation on 50.

13. *McGraw-Hill Modern Scientists and Engineers*, s.v. "Starr, Victor Paul"; and Abraham H. Oort, "Angular Momentum Cycle in the Atmosphere–Ocean–Solid Earth System," *Bulletin of the American Meteorological Society* 70 (1989): 1231–42, esp. 1231–32. For insights to academic meteorology in the 1940s, see Horace R. Byers, "The Founding of the Institute of Meteorology at the University of Chicago," *Bulletin of the American Meteorological Society* 57 (1976): 1343–45. The suggestion of logrolling is mine.

14. Starr, *Basic Principles of Weather Forecasting*, 51.

15. Lawrence A. Hughes, "A Note on the Effect of the Great Lakes," *Weather Bureau Topics* 16: 242–43. For further information on Hughes's life and work, see Dan Smith and Bob Glahn, "Lawrence A. Hughes: 1918–2008," *Bulletin of the American Meteorological Society* 89 (2008): 1188–90.

16. H. C. Willett, "The Forecast Problem," in *Compendium of Meteorology*, ed. Thomas F. Malone, 731–46 (Boston: American Meteorological Society, 1951), quotation on 731. Historian Frederik Nebeker, whose book *Calculating the Weather* brought Willett's assessment to my attention, underscored the point with a statistical chart showing an average error in thirty-hour predictions of surface pressure of roughly 64 percent from 1948 through the mid-1950s. See Frederik Nebeker, *Calculating the Weather: Meteorology in the 20th Century* (San Diego: Academic Press, 1995), 172. The chart was reproduced from George P. Cressman, "Dynamic Weather Prediction," in *Meteorological Challenges: A History*, ed. D. P. Mcintyre, 179–207 (Ottawa: Information Canada, 1972), chart on 185.

17. Cressman, "Dynamic Weather Prediction," 185.

18. Willett, "The Forecast Problem," 738.

19. J. G. Charney, "Dynamic Forecasting by Numerical Process," in Willett, *Compendium of Meteorology*, 470–82; quotations on 470.

20. Lewis F. Richardson, *Weather Prediction by Numerical Process* (Cambridge: Cambridge Univ. Press, 1922), esp. 181–213. For background and critique, see Nebeker, *Calculating the Weather*, 58–82; and Peter Lynch, *The Emergence of Numerical Weather Prediction: Richardson's Dream* (Cambridge: Cambridge Univ. Press, 2006).

21. Charney, "Dynamic Forecasting by Numerical Process," 470. Decades later, a computer program that replicated Richardson's model vindicated his manual calculations, expedited with slide rule and log tables, thus laying the blame for the failed forecast

on deficiencies in both his model and his data. See Lynch, *The Emergence of Numerical Weather Prediction*, 119, 243–46.

22. Lynch, *The Emergence of Numerical Weather Prediction*, 479. Also see George W. Platzman, "The ENIAC Computations of 1950—Gateway to Numerical Prediction," *Bulletin of the American Meteorological Society* 60 (1979): 302–12.

23. Norman A. Phillips, "Jule Charney's Influence on Meteorology," *Bulletin of the American Meteorological Society* 63 (1982): 492–98; quotation on 493. Also see Nebeker, *Calculating the Weather*, 135–51.

24. Edward N. Lorenz, "Reflections on the Conception, Birth, and Childhood of Numerical Weather Prediction," *Annual Review of Earth and Planetary Science* 34 (2006): 37–45.

25. Norman A. Phillips, "A Simple Three-Dimensional Model for the Study of Large-Scale Extra-Tropical Flow Patterns." *Journal of Meteorology* 8 (1951): 381–94.

26. J. G. Charney and N. A. Phillips, "Numerical Integration of the Quasi-Geostrophic Equations for Barotropic and Simple Barotropic Flows," *Journal of Meteorology* 10 (1953): 71–99; quotations on 71.

27. The NCEP graph in figure 3.4 uses a modified skill score calculated as $100 \times (1 - S_1/70)$, where S_1 is the Teweles-Wobus skill score, introduced in 1954. Because a forecast with a lower S_1 is more reliable than a forecast with a higher S_1, the NCEP transformed the raw S_1 estimates so that an increase would now represent an improvement. Division by 70 reflects recognition that a forecast with an S_1 of 70 is "worthless"—no better than a prediction based on climatological data for the day in question—so that the value 0 on the new scale thus reflects a worthless forecast. Because a forecast with an S_1 of 20 is considered "extremely good" or "near perfect," the corresponding transformed value of 71.4, surpassed in the late 1990s for the thirty-six-hour forecast, represents a significant achievement. For discussion of S_1, see Frederick G. Shuman, "History of Numerical Weather Prediction at the National Meteorological Center," *Weather and Forecasting* 4 (1989): 286–96; and Panel on the Road Map for the Future, National Weather Service, and National Weather Service Modernization Committee, National Research Council, *A Vision for the National Weather Service: Road Map for the Future* (Washington, DC: National Academies Press, 1999), esp. 30–32; quotations on 31.

28. Ensemble modeling became an operational component of NOAA weather prediction in December 1992. Eugenia Kalnay, Stephen J. Lord, and Ronald D. McPherson, "Maturity of Operational Numerical Weather Prediction: Medium Range," *Bulletin of the American Meteorological Society* 79 (1998): 2753–69; and Eugenia Kalnay, *Atmospheric Modeling, Data Assimilation and Predictability* (Cambridge: Cambridge Univ. Press, 2003), 227–49.

29. Lawrence A. Hughes, "Precipitation Probability Forecasts—Problems Seen via a Comprehensive Verification," *Monthly Weather Review* 107 (1979): 129–39.

30. Verification was still an important management tool at the end of the twentieth century. See Gary Alan Fine, *Authors of the Storm: Meteorologists and the Culture of Prediction* (Chicago: Univ. of Chicago Press, 2007), 178–94.

31. I say *his* because the forecaster cadre of the late 1970s included many former military meteorologists and few women. Though female weather forecasters are not uncommon in the early twenty-first century, operational meteorology remains strongly gendered, perhaps because women are averse—or are expected to be averse—to rotating shifts. See Fine, *Authors of the Storm*, 39.

32. Hughes, "Precipitation Probability Forecasts," 138.

33. Ibid., 134 (his emphasis).

34. The twelve journals are the *Journal of the Atmospheric Sciences* (covered since volume 1, 1944), the *Journal of Applied Meteorology and Climatology* (since volume 1, 1962), the *Journal of Physical Oceanography* (since volume 1, 1971), the *Monthly Weather Review* (since issue 5 of "volume 0," for May 1872, and volume 1, 1873), the *Journal of Atmospheric and Oceanic Technology* (since volume 1, 1984), *Weather and Forecasting* (since volume 1, 1986), the *Journal of Climate* (since volume 1, 1988), the *Journal of Hydrometeorology* (since volume 1, 2000), the *Bulletin of the American Meteorological Society* (since volume 51, 1970), *Meteorological Monographs* (since volume 29, 2003), *Weather, Climate, and Society* (since volume 1, 2009), and *Earth Interactions* (since volume 1, 1997). The searchable database is available at American Meteorological Society, AMS Journals Online, http://journals.ametsoc.org/. My data are admittedly deficient because the twelve journals in the AMS database are an incomplete subset of what's been published. My search was further handicapped because the database did not include pre-1970 issues of the Society's *Bulletin*, initiated in 1920, and thus ignored John Remick's possibly pioneering use of the term in January 1942. Remick used the term, unhyphenated, as adjective and noun, not as a compound adjective modifying "snow." See Remick, "The Effect of Lake Erie on the Local Distribution of Precipitation in Winter (I)," 1.

35. S. Petterssen and P. A. Calabrese, "On Some Weather Influences Due to Warming of the Air by the Great Lakes in Winter," *Journal of Meteorology* 16 (1959): 646–52; quotations on 646 and 652.

36. R. L. Peace, Jr., and R. B. Sykes, Jr., "Mesoscale Study of a Lake Effect Snow Storm," *Monthly Weather Review* 94 (1966): 495–507; quotation on 506.

37. Tom Niziol and Dave Eichorn, "Robert B. Sykes, Jr., 1917–1999," *Bulletin of the American Meteorological Society* 80 (1999): 1170–71; quotation on 1170.

38. B. L. Wiggin, "Great Snowstorms of the Great Lakes," *Weatherwise* 3 (1950): 123–26; quotation on 124. Bernard Wiggin was meteorologist in charge at the Weather Bureau's Buffalo office for many years. The term also appeared in Ernest C. Johnson and Conrad P. Mook, "The Heavy Snowstorm of January 28–30, 1953, at the Eastern End of Lake Ontario." *Monthly Weather Review* 81 (1953): 26–30; quotation on 26. Sykes distinguished a snowburst from both a typical blizzard and a blizzard-burst in Robert B.

Sykes, Jr., "Scales and Patterns Relating to Lake Effect Snow Situations off Eastern Lake Ontario," *Proceedings of the 35th Eastern Snow Conference*, Hanover, NH, February 2–3, 1978, 2–3.

39. National Science Foundation, "Inside a Snowstorm: Scientists Obtain Close-Up Look at Old Man Winter," press release 11-004, January 11, 2011.

40. Robert B. Sykes, Jr., "Oswego's Tardy, Tough Winter of 1971–72," *Weatherwise* 6 (1972): 276–83; and Robert B. Sykes, Jr., "The 1972 Blizzard-Burst at the Eastern Snow Conference in Oswego," *Proceedings of the 30th Eastern Snow Conference*, Amherst, NY, February 8–9, 1973, 120–24.

41. James E. Jiusto and Michael L. Kaplan, "Snowfall from Lake-Effect Storms," *Monthly Weather Review* 100 (1972): 62–66; quotation on 62.

42. James Edward Jiusto, "Nucleation Factors in the Development of Clouds" (PhD diss., Pennsylvania State Univ., 1968); and Ronald L. Lavoie, "A Mesoscale Numerical Model and Lake-Effect Storms" (PhD diss., Pennsylvania State Univ., 1968).

43. Ronald L. Lavoie, "A Mesoscale Numerical Model of Lake-Effect Storms," *Journal of the Atmospheric Sciences* 29 (1972): 1025–40; quotations on 1038 and 1039.

44. Thomas A. Niziol, "Operational Forecasting of Lake Effect Snowfall in Western and Central New York," *Weather and Forecasting* 2 (1987): 310–21; quotation on 321. For a picture of forecasting practices at Buffalo a decade earlier, see Benjamin Kolker, "Current Forecast Procedures for Lake Effect Snows in Western New York Especially Related to 1976–1977 and 1977–1978 Winters," *Proceedings of the 35th Eastern Snow Conference*, Hanover, NH, February 2–3, 1978, 17–35.

45. Thomas A. Niziol, Warren R. Snyder, and Jeff S. Waldstreicher, "Winter Weather Forecasting Throughout the Eastern United States. Part IV: Lake Effect Snow," *Weather and Forecasting* 10 (1995): 61–77; quotations on 66 and 75.

46. Timothy D. Crum and Ron L. Alberty, "The WSR-88D and the WSR-88D Operational Support Facility," *Bulletin of the American Meteorological Society* 74 (1993): 1669–87.

47. Elbert W. Friday, Jr., "The Modernization and Associated Restructuring of the National Weather Service: An Overview," *Bulletin of the American Meteorological Society* 75 (1994): 43–52. The count of 230 local offices is from Fine, *Authors of the Storm*, 214. According to Friday, the plan called for 116 WFOs, but Fine counted 122. An NWS list dated May 18, 2006, reports 123 WFOs, including one each in American Samoa, Guam, Hawaii, and Puerto Rico, and three in Alaska. See http://www.weather.gov/mirs/public/prods/reports/xl/cwfa_square_miles_by_area.xls.

48. Fine, *Authors of the Storm*, 214.

49. Ibid., 32.

50. Thomas Niziol, e-mail communication, June 30, 2011.

51. NEXRAD Panel, National Weather Service Modernization Committee, National Research Council, *Toward a New National Weather Service: Assessment of*

NEXRAD *Coverage and Its Associated Weather Services* (Washington, DC: National Academies Press, 1995), esp. 8–32.

52. National Oceanic and Atmospheric Administration, Office of the Federal Coordinator for Meteorological Services and Supporting Research, *Doppler Radar Meteorological Observations, Part A: System Concepts, Responsibilities, and Procedures*, FCM-H11A-2009 (Washington, DC, May 2009), table 3-2 on page 3-9.

53. Rodger A. Brown et al., "Improved Detection Using Negative Elevation Angle for Mountaintop WSR-88Ds. Part III: Simulations of Shallow Convective Activity over and around Lake Ontario," *Weather and Forecasting* 22 (2007): 839–52; quotations on 840 and 851.

54. Robert J. Serafin and James W. Wilson, "Operational Weather Radar in the United States: Progress and Opportunity," *Bulletin of the American Meteorological Society* 81 (2000): 501–18, esp. 504. A "glimmer of hope" appeared in 2010, when the National Weather Service installed a new radar with lower elevation angles on the Pacific coast in Washington State. Rodger A. Brown, National Severe Storms Laboratory, Norman, OK, e-mail communication, July 27, 2010.

55. SRI International, *Supplemental Environmental Assessment (SEA) of the Effects of Electromagnetic Radiation from the WSR-88D Radar: Final Report Prepared for the Next Generation Weather Radar Joint System Program Office*, April 1993; and Timothy D. Crum, NOAA Radar Operations Center, Norman, OK, e-mail communication, June 29, 2011.

56. US General Accounting Office, *National Weather Service: Modernization Activities Affecting Northwestern Pennsylvania*, Report no. GAO/AIMD-97-156, September 1997; quotation on 24.

57. A smaller, less powerful radar run by the Federal Aviation Administration was eventually installed at the local airport. See Gary Wesman, "Bill Calls for Better Plan to Predict Lake-Effect Snow," *Erie Times-News*, November 16, 2001.

58. George Miller, "WJET Acquires Doppler Radar to Forecast Weather," *Erie Times-News*, December 2, 1998; and Dave Richards, "Tonight's Top Story: Sunny but Cool, High near 45 Degrees," *Erie Times-News*, December 2, 1998.

59. Thomas Atkins, telephone conversation, July 26, 2010.

60. Edward A. Mahoney and Thomas A. Niziol, "BUFKIT: A Software Application Toolkit for Predicting Lake-Effect Snow," *Preprints, 13th International Conference on Interactive Information and Processing Systems (IIPS) for Meteorology, Oceanography, and Hydrology*, February 2–7, 1997, Long Beach, CA, American Meteorological Society, 388–91; and Thomas A. Niziol and Edward A. Mahoney, "The Use of High Resolution Hourly Forecast Soundings for the Prediction of Lake-Effect Snow," *Preprints, 13th International Conference on Interactive Information and Processing Systems (IIPS) for Meteorology, Oceanography, and Hydrology*, February 2–7, 1997, Long Beach, CA, American Meteorological Society, 92–95.

61. The most up-to-date source is the BUFKIT page on the Buffalo WFO's Web page, http://www.wbuf.noaa.gov/bufkit/bufkit.html. For additional information, see the NWS Buffalo Mesoscale Model Page, http://www.erh.noaa.gov/buf/mm/mesomodels. html. For an example of a research study that used BUFKIT, see Robert J. Ballentine, Alfred J. Stamm, and Eugene E. Chermack, "Mesoscale Model Simulation of the 4–5 January 1995 Lake-Effect Snowstorm," *Weather and Forecasting* 13 (1998): 893–920.

62. National Weather Service, Buffalo Forecast Office, Aphid storm web page, http://www.erh.noaa.gov/buf/storm101206.html.

63. Andrew Freedman, "Anatomy of a Forecast: 'Armageddon' Takes Buffalo by Surprise," *Weatherwise* 60 (July/August 2007): 16–21; quotation on 19.

64. Robert S. Hamilton, David Zaff, and Thomas Niziol, "A Catastrophic Lake Effect Storm over Buffalo, NY, October 12–13, 2006," *Preprint, 22nd Conference on Weather Analysis and Forecasting/18th Conference on Numerical Weather Prediction*, June 25–29, 2007, Park City, UT, paper 3A.6; quotation on 6.

65. Gary M. Lackmann, "Analysis of a Surprise Western New York Snowstorm," *Weather and Forecasting* 16 (2001): 99–116; quotation on 101.

66. Yarice Rodriguez, David A. R. Kristovich, and Mark R. Hjelmfelt, "Lake-to-Lake Cloud Bands: Frequencies and Locations," *Monthly Weather Review* 135 (2007): 4202–13; quotation on 4202. Also see Robert J. Ballentine et al., "Mesoscale Model Simulation of the 4–5 January 1995 Lake-Effect Snowstorm," *Weather and Forecasting* 13 (1998): 893–920; and Greg E. Mann, Richard B. Wagenmaker, and Peter J. Sousounis, "The Influence of Multiple Lake Interactions upon Lake-Effect Storms," *Monthly Weather Review* 130 (2002): 1510–30.

67. The conference's agenda and PowerPoint presentations are online at Cornell University's Earth and Atmospheric Sciences website, at http://glomw.eas.cornell.edu/. I presented one of the papers, a somewhat condensed version of chapter 2.

68. Harry R. Glahn and David P. Ruth, "The New Digital Forecast Database of the National Weather Service," *Bulletin of the American Meteorological Society* 84 (2003): 195–201; and National Weather Service, National Digital Forecast Database, http://www.weather.gov/ndfd/.

69. Thomas Niziol, e-mail communication, June 30, 2011.

70. National Weather Service, Buffalo Forecast Office, "Lake Effect Storm Alewife–November 26–27, 2010," http://www.erh.noaa.gov/buf/lakeeffect/lake1011/a/stormsuma.html.

71. National Weather Service, Marquette Forecast Office, "Probabilistic Precipitation/SnowAmt Detailed Definition (taken from NOAA/NWS Detroit/Pontiac)," http://www.crh.noaa.gov/mqt/?n=pqpf_explain.

72. On this day the 12-hour period ended at 7 a.m., not 8 a.m.

73. For discussion of the sociological implications of NWS modernization, see Fine, *Authors of the Storm*, esp. 163–70. For discussion of the social and cognitive implications,

see Phaedra Daipha, "Visual Perception at Work: Lessons from the World of Meteorology," *Poetics* 38 (2010): 151–65.

74. The NDFD and its associated software package, IFPS (Interactive Forecast Preparation System), inspired a searching discussion of the role of human forecasters. See Clifford E. Mass, "IFPS and the Future of the National Weather Service," *Weather and Forecasting* 18 (2003): 75–79; Bob Glahn, "Comments on 'IFPS and the Future of the National Weather Service,'" *Weather and Forecasting* 18 (2003): 1299–1304; Neil A. Stuart et al., "The Future of Humans in an Increasingly Automated Forecast Process," *Bulletin of the American Meteorological Society* 87 (2006): 1497–1502; and David M. L. Sills, "On the MSC Forecasters Forums and the Future Role of the Human Forecaster," *Bulletin of the American Meteorological Society* 90 (2009): 619–27.

4. Impacts

1. Charles McChesney, "Syracuse Accolade: 'Worst Winter City,'" *Syracuse Post-Standard*, September 10, 2010.

2. Mark Weiner, "Why We Can't Stop Talking about Weather," *Syracuse Post-Standard*, September 11, 2001; and "Consider This," *Syracuse Post-Standard*, September 12, 2010.

3. Jeffrey B. Halverson and Thomas D. Rabenhorst, "Mega-Snow in the Megalopolis: The Mid-Atlantic's Blockbuster Winter of 2009–2010," *Weatherwise* 63 (July/August 2010): 16–22.

4. The *Syracuse Post-Standard* could not resist the opportunity to point out that, as of mid-February, Baltimore's total snowfall (at the airport) exceeded Syracuse's. See Charles McChesney, "Baltimore Snow Total Tops Syracuse's," *Syracuse Post-Standard*, February 12, 2010.

5. Northeast Regional Climate Center, NOWData—NOAA Online Weather Data, http://www.nrcc.cornell.edu/page_nowdata.html.

6. National Weather Service Glossary, http://www.weather.gov/glossary/.

7. Robert W. Schwartz and Thomas W. Schmidlin, "Climatology of Blizzards in the Conterminous United States, 1959–2000," *Journal of Climate* 15 (2002): 1765–72.

8. Ibid., 1771. Their study excluded Alaska, home to some of the most brutal weather in North America.

9. Stanley A. Changnon, David Changnon, and Thomas R. Karl, "Temporal and Spatial Characteristics of Snowstorms in the Contiguous United States," *Journal of Applied Meteorology and Climatology* 45 (2006): 1141–55.

10. Developed in 1971 by University of Chicago meteorologist Theodore Fujita (1920–1998), the Fujita scale uses after-the-fact assessments of structural damage to estimate wind velocity in tornadoes, which often strike with little warning. (The "Enhanced Fujita Scale" was introduced in 2006 to provide more consistent estimates of wind

speed.) Because wind speed in hurricanes can be estimated before a storm strikes, the Saffir-Simpson scale, which is based on studies by structural engineer Herbert Saffir (1917–2007) and meteorologist Robert Simpson (1912–), plays an important role in alerting emergency managers and the public to the likely consequences of these vast but relatively slow moving storms. For further discussion, see James R. McDonald, "T. Theodore Fujita: His Contribution to Tornado Knowledge Through Damage Documentation and the Fujita Scale," *Bulletin of the American Meteorological Society* 82 (2001): 63–72; Robert H. Simpson, "The Hurricane Disaster Potential Scale," *Weatherwise* 27 (1974): 169–86; and Robert Simpson and Herbert Saffir, "Tropical Cyclone Destructive Potential by Integrated Kinetic Energy," *Bulletin of the American Meteorological Society* 88 (2007): 1799–1800. Also see H. Michael Mogil, *Extreme Weather: Understanding the Science of Hurricanes, Tornadoes, Floods, Heat Waves, Snow Storms, Global Warming and Other Atmospheric Disturbances* (New York: Black Dog & Leventhal, 2007), esp. 20, 52–55, and 117–18.

11. Paul J. Kocin and Louis W. Uccellini, "A Snowfall Impact Scale Derived from Northeast Storm Snowfall Distributions," *Bulletin of the American Meteorological Society* 85 (2004): 177–94.

12. As part of the calculation, a geographic information system is used to estimate population within each of four overlapping snowfall categories (>4 in., >10 in., >20 in., and >30 in.). For detail on the calculation of NESIS scores, see Michael F. Squires and Jay H. Lawrimore, "Development of an Operational Northeast Snowfall Impact Scale," *Proceedings of the 22nd Conference on Interactive Information and Processing Systems (IIPS) for Meteorology, Oceanography, and Hydrology,* January 29–February 2, 2006, Atlanta, GA. Online at the National Climatic Data Center, http://www.ncdc.noaa.gov/snow-and-ice/docs/squires.pdf.

13. National Climate Data Center, The Northeast Snowfall Impact Scale (NESIS), http://www.ncdc.noaa.gov/snow-and-ice/nesis.php.

14. Because they were so close in time—February 4–7 and 9–11—these distinct weather systems had the combined impact of a single snowstorm with a NESIS score of 11.60, sufficient for Category 5. See Jeffrey B. Halverson and Thomas D. Rabenhorst, "Mega-Snow in the Megalopolis: The Mid-Atlantic's Blockbuster Winter of 2009–2010," *Weatherwise* 63 (July/August 2010): 16–22. Combined score provided by David A. Call, e-mail communication, June 20, 2011.

15. National Climatic Data Center, "2009/2010 Cold Season," May 24, 2010, http://www.ncdc.noaa.gov/special-reports/2009-2010-cold-season.html.

16. Quotations from National Weather Service, Buffalo, "Lake Effect Storm 'Amaryllis,' December 10–12, 2009," http://www.erh.noaa.gov/buf/lakeffect/lake0910/a/storms uma.html.

17. Donn Esmonde, "Bright Side of the Season's First Storm," *Buffalo News*, December 11, 2009.

18. Capsule summaries described the three-flake Eskimo storm (January 25–27, 2000), on the margin of a NESIS Category 2 (January 24–26) northeast snowstorm, as a "hybrid synoptic/lake effect storm," and the two-flake Mealworm storm (February 15–16, 2007), vaguely within the fringe of a NESIS Category 3 (February 12–15) storm stretching from Colorado to Maine, as a "post-synoptic relatively brief event." By contrast, no connection was noted between the two-flake Korat storm (February 13–14, 2006) and a slightly earlier Category 3 (February 12–13) NESIS event, the map of which included a weak outlier east of Lake Erie. Korat was a comparatively brief storm, and relatively late in the season because Lake Erie was still largely unfrozen. Quotations are from seasonal summaries on the NWS Lake Effect Page, http://www.erh.noaa.gov/buf/lakepage.php, which also provided the maps discussed. Maps of the NESIS storms are online at the NESIS web page, http://www.ncdc.noaa.gov/snow-and-ice/nesis.php.

19. A map reporting the single month with the greatest number of snowstorms shows December as the peak month for the Upper Peninsula of Michigan as well as for a narrow belt along Lakes Erie and Ontario and the Canadian border, and extending from northeastern Ohio into northern Maine. February is the peak month for the Appalachians from Pennsylvania into northern Georgia, and for a narrow strip along the Atlantic coast from North Carolina through Maine. January is the peak month for the rest of the Northeast. Changnon, Changnon, and Karl, "Temporal and Spatial Characteristics of Snowstorms," 1153. As I note in chapter 6, evidence suggests that lake-effect snow is extending later in the season.

20. William B. Meyer, *Americans and Their Weather* (New York: Oxford Univ. Press, 2000), 35–36.

21. Bernard Mergen, *Snow in America* (Washington, DC: Smithsonian Institution Press, 1997), 53.

22. Ibid., 41–50. For discussion of present-day challenges, see Stanley A. Changnon, *Railroads and Weather: From Fogs to Floods and Heat to Hurricanes, the Impacts of Weather and Climate on American Railroading* (Boston: American Meteorological Society, 2006), esp. 15, 76–82, 96–99.

23. Numbers from William D. Middleton, *Metropolitan Railways: Rapid Transit in America* (Bloomington, IN: Indiana Univ. Press, 2003), 2; and John H. White, Jr., "Horse Power," *American Heritage of Invention and Technology* 8 (Summer 1992): 41–51. Middleton reported the annual revenue for 1881 as "in the vicinity of $1.25 billion," which is clearly an overstatement—passengers would never pay a dollar a ride.

24. American Public Works Association, Committee on Street Cleaning, *Street Cleaning Practice* (Chicago: American Public Works Association, 1938), 188.

25. Blake McKelvey, *Snow in the Cities: A History of America's Urban Response* (Rochester, NY: Univ. of Rochester Press, 1995), 110.

26. Rochester Bureau of Municipal Research, *Report on the Problem of Snow Removal in the City of Rochester, N.Y.* (Rochester, 1917); quotations on 24.

27. Blake McKelvey, "Snowstorms and Snow Fighting—The Rochester Experience," *Rochester History* 27 (1965): 1–24; esp. 20.

28. Rochester Bureau of Municipal Research, *Report on the Problem of Snow Removal*, 20–21.

29. City of Rochester, New York, "Sidewalk Snow Removal," http://www.cityof rochester.gov/.

30. Hamilton's mayor, Sue McVaugh, confirmed that the village still plows residential sidewalks (e-mail October 25, 2010). According to public works director Sean Graham, the village uses a variety of small tractors to plow sidewalks (telephone conversation, October 28, 2010).

31. "Snowbelt City Plows 300 Miles of Sidewalks," *Public Works* 106 (August 1975): 65; and Kristen Johnson, "Jamestown Holding Its Own Against Snow," *Jamestown Post-Journal*, January 5, 2010.

32. See, for example, Dan McGuire, "City Balks at Shoveling Sidewalks," *Syracuse Post-Standard*, February 21, 1990; and Brian Meyer, "Buffalo Officials Discussing Sidewalk Plowing Service," *Buffalo News*, February 15, 2009.

33. For evidence of increased recognition of parked vehicles as an impediment to snow removal, compare American Public Works Association, Committee on Street Cleaning, *Street Cleaning Practice* (Chicago, 1938), 204; American Public Works Association, Street Sanitation Committee, *Street Cleaning Practice*, 2nd ed. (Chicago, 1959), 201–2; and Rodney E. Fleming, ed., *Street Cleaning Practice*, 3rd ed. (Chicago: American Public Works Association, 1978), 378–80.

34. See, for example, Liz Kay, "Baltimore Snow Emergency Routes Still Need to Be Cleared," *Baltimore Sun*, February 9, 2010.

35. David A. Call, "Urban Snow Events in Upstate New York: An Integrated Human and Physical Geographical Analysis" (master's thesis, Syracuse Univ., 2004), 69–74.

36. David A. Call, "Rethinking Snowstorms as Snow Events," *Bulletin of the American Meteorological Society* 86 (2005): 1783–93; quotations on 1787.

37. David A. Call, telephone conversation, June 20, 2011.

38. "What the Various States Are Doing in the Way of Snow Removal," *Roads and Streets* 68 (1928): 545–60; quotations on 545 and 547.

39. "How the Various States Are Handling Their Snow Removal Problems," *Roads and Streets* 69 (1929): 378–400; quotation on 386.

40. "Snow Removal That Is Snow Removal," *Roads and Streets* 82 (August 1939): 31. Houghton County had 900 miles of county roads. Its highway department plowed 600 miles of the county's own roads plus about 100 miles of state highway, under contract, for a total of 700 miles.

41. Carl F. Winkler, "Plowing Methods for Very Heavy Snowfall," *Roads and Streets* 93 (September 1950): 59–62.

42. B. C. Tiney, "Winter Maintenance in Michigan," *Public Works* 56 (1925): 342–44,

43. Squire E. Fitch, "Snow Removal in Chautauqua County, New York," *Roads and Streets* 75 (October 1933): 72–74; quotation on 74.

44. Squire E. Fitch, "Open Roads in New York's Snow Belt," *Public Works* 69 (November 1938): 12–14; quotation on 12.

45. For examples, see Richard S. Cortina and Thomas A. Low, "Development of New Routes for Snow and Ice Control," *Public Works* 132 (August 2001): 20–23; and Robert A. Ferlis, Shabed Rowshan, and Cathy Frye, "Safe Plowing: Applying Intelligent Vehicle Technology," *Public Roads* 64 (January/February 2001): 3–8.

46. Frank F. Harmon, "How We Get 'Florida' Streets in the Winter," *Public Works* 84 (August 1953): 72–74; quotations on 72 and 74.

47. Actual allocated expenses listed in the 2011 Budget Summary (dated November 9, 2011), provided by Angela Epolito, assistant to the supervisor, Town of DeWitt, NY.

48. New York State Department of Transportation, Office of Transportation Maintenance, "Snow and Ice Control," https://www.dot.ny.gov/divisions/operating/oom/transportation-maintenance/snow-and-ice.

49. Robert T. Carrier, "County Makes Year-Round Preparations for Battle with Snow," *Public Works* 87 (September 1956): 117–19.

50. Rick Moriarty, "County Exec Told: Cut More—Legislators Look for Ways to Reduce Budget, Tax Rate," *Syracuse Post-Standard*, September 16, 2010.

51. See, for example, Robert A. Muller, "An Analysis of Factors Contributing to the Costs of Highway Snow Removal in Oswego County, New York," *Proceedings of the 17th Eastern Snow Conference*, Troy, NY, February 4–5, 1959, 129–39.

52. G. R. Thompson, "Snow Removal in Detroit," *Public Works* 57 (1926): 410–12; quotation on 411.

53. Harmon, "How We Get 'Florida' Streets in the Winter," 72.

54. National Research Council, Committee on Progress and Priorities of US Weather Research and Research-to-Operations Activities, *When Weather Matters: Science and Service to Meet Critical Societal Needs* (Washington, DC: National Academies Press, 2010), 90.

55. I have a strong hunch that any substantial expansion of NWS service to municipalities would encounter strong political opposition from private-sector forecasting firms.

56. National Research Council, *When Weather Matters*, 52–53.

57. David A Kuemmel, *Managing Roadway Snow and Ice Control Operations* (Washington, DC: National Academy Press, 1994), 19–23.

58. Jeff Archer, "Districts Scramble to Make Up for Lost Time," *Education Week* 15 (February 28, 1996): 5.

59. Thomas W. Schmidlin, "Impacts of Severe Winter Weather During December 1989 in the Lake Erie Snowbelt," *Journal of Climate* 6 (1993): 759–67; quotation on 759. Of the 52 surveys mailed, only 39 were returned.

60. See, in particular, Ellen R. Delisio, "To Close or Not to Close: A Superintendent's Winter Worry," EducationWorld.com, February 3, 2004 (updated September 24, 2008); and P. Susan Mamchak and Steven R. Mamchak, *Complete School Communications Manual, with Sample Letters, Forms, Bulletins, Policies, and Memos* (Englewood Cliffs, NJ: Prentice-Hall, 1984), 127–28.

61. Jack Turcotte, "Anatomy of My Snow Daze," *School Administrator* 60 (February 2003), 29.

62. Randy L. Dewar, "The Snow Day: One Tough Call," *School Administrator* 60 (February 2003): 26–28.

63. Wooster City School District, "Snow Day—How Is the Decision Made?" http://www.woostercityschools.org/district/content-page/school-closing-information.

64. National Weather Service, Buffalo Forecast Office, "Lake Effect Storm Carp—December 5–8, 2010," http://www.erh.noaa.gov/buf/lakeeffect/lake1011/c/stormsumc.html.

65. Making a real map of school closings proved more difficult than forming a mental image. State education officials haven't published a map of district boundaries, and the detailed electronic school districts map from the New York State GIS Clearinghouse is cluttered with jagged boundaries that follow property lines, rather than town or county boundaries. Present-day school districts evolved from successive consolidations of rural schools during the first half of the last century. Rural districts have been allowed to exchange property and students with neighboring jurisdictions, and a few prominent outliers suggest that contiguity is not a constraint on educational gerrymandering. Avoiding clutter and confusion (mine as well as yours) called for a heavy dose of cartographic generalization.

66. Henry Bothwell, telephone interview, January 24, 2011. According to the district website, NICE is an acronym for several townships: National Mine, Ishpeming, Champion, and Ely. The rural school district was formed from the consolidation of these four districts in the late 1960s and early 1970s. NICE Community School District, http://www.nice.k12.mi.us/.

67. Phillip B. Lee, "Transportation Versus Snow," *Proceedings of the 23rd Eastern Snow Conference*, Hartford, CT, February 4–5, 1959, 80–87.

68. National Weather Service, Buffalo Forecast Office, "Lake Effect Storm Bluegill—December 1–3, 2010," http://www.erh.noaa.gov/buf/lakeeffect/lake1011/b/stormsumb.html.

69. Cathy Woodruff, "Thruway Drops Tolls Near Buffalo," *Albany Times Union*, October 31, 2006; Robert J. McCarthy, "Motorists Livid over Thruway Strandings," *Buffalo News*, December 3, 2010; Carolyn Thompson, "Furious Motorists Admonish Thruway Officials for Snow Delays," Democrat&Chronicle.com, December 4, 2010; and Carolyn Thompson, "NY Road Officials Promise Quicker Response to Snow," *Buffalo News*, December 16, 2010.

70. David Zaff, "Lake Effect Bands and Highways Don't Mix," Great Lakes Operational Meteorology Workshop, March 21–23, 2011, Ithaca, NY, presentation online at http://glomw.eas.cornell.edu/.

71. For insights on how snowgates work, see Gregory E. Thompson and Matthew J. Via, Automatic Gate Closing System for Freeway Interchanges, US Patent 6,696,977, filed April 10, 2002, and issued February 24, 2004.

72. Mark Monmonier and Alberto Giordano, "GIS in New York State County Emergency Management Offices: User Assessment," *Applied Geographic Studies* 2 (1998): 95–109; and FEMA Memorandum from Gil Jameson, Chair, Regional Risk Assessment Group, June 21, 1995 (cited in Monmonier and Giordano).

73. New York State Office of Emergency Management, *New York State Standard Multi-Hazard Mitigation Plan*, esp. table 4-2, Recommended Project Types by Natural Hazards, on page 4–13. The entire plan is online at http://www.semo.state.ny.us/programs /planning/hazmitplan.cfm.

5. Records

1. Charles A. Sauer, "Snowy Syracuse Still Whitest of Them All," *Syracuse Post-Standard*, January 3, 1990; and Sean Kirst, "To the Victor Goes the Snow . . . Trophy," *Syracuse Post-Standard*, March 1, 2002. Quotation from Peter Chaston, telephone conversation, May 11, 2011.

2. William Kates, "Forecasters Suggest Resurrection of New York's Snowfall Competition," *USA Today*, January 16, 2003; William Kates, "Syracuse's Snowball Buries the Competition Once Again," Associated Press State and Local Wire, May 5, 2004; and Sean Kirst, "An Idea That Just Snowballed," *Syracuse Post-Standard*, Neighbors Syracuse section, November 11, 2005.

3. Sean Kirst, "To the Victor Goes the Snow . . . Trophy." Alexander, who was mayor from 1970 to 1985, was convicted of racketeering and tax evasion and ended his political career at a federal prison in Wisconsin.

4. DeCoursey thanked Buffalo resident Steve Madsen for the data that he posted online at Comparison Golden Snowball City Stats, http://www.goldensnowball.com /yearly-winners-golden-award.htm. I found a more up-to-date listing on Wikipedia, at http://en.wikipedia.org/wiki/Snowfall_Statistics_for_Golden_Snowball_Award_Cities.

5. Associated Press State and Local Wire, "Syracuse Plows Its Way to Win Snowball Race," May 9, 2005.

6. Nolan J. Doesken and Arthur Judson, *The Snow Booklet: A Guide to the Science, Climatology, and Measurement of Snow in the United States* (Fort Collins: Colorado Climate Center, Dept. of Atmospheric Science, Colorado State Univ., 1996; 2nd ed., 1997), 44.

7. Quoted in Sean Kirst, "The Lament of Silver Lake, Colo.: Say It Ain't Snow," *Syracuse Post Standard*, January 16, 1997.

8. Ibid.

9. C. F. Brooks, "On Maximum Snowfalls," *Bulletin of the American Meteorological Society* 19 (1938): 87.

10. J. L. H. Paulhus, "Record Snowfall of April 14–15, at Silver Lake, Colorado," *Monthly Weather Review* 81 (1953): 38–40.

11. National Weather Service, Office of Meteorology, *Evaluation of the Reported January 11–12, 1997, Montague, New York, 77-inch, 24-hour Lake-effect Snowfall* (Silver Spring, MD: National Oceanic and Atmospheric Administration, 1997).

12. Ibid., ES-2.

13. "WSFO BUF's Snow Spotter Training Documentation," quoted in National Weather Service, *Evaluation of the Montague Snowfall*, A-1.

14. National Weather Service, *Evaluation of the Montague Snowfall*, 39.

15. Ibid., 30.

16. Ibid., 12.

17. Ibid., A-1.

18. Ibid., 18.

19. Mark Weiner, "Montague, Known for Snow, Gets Doppler Radar," *Syracuse Post-Standard*, March 27, 1998.

20. National Weather Service, *Evaluation of the Montague Snowfall*, 4–6.

21. Ibid., 7, 39. For activation procedures, record investigation reports, and related materials, see the National Climate Extremes Committee website, http://www.ncdc .noaa.gov/extremes/ncec/.

22. Ibid., 34–35, 38, and appendix B.

23. Ibid., 38.

24. Ibid., 38–39.

25. Robert Cifelli et al., "The Community Collaborative Rain, Hail, and Snow Network: Informal Education for Scientists and Citizens," *Bulletin of the American Meteorological Society* 86 (2004): 1069–77; and Community Collaborative Rain, Hail and Snow Network, http://www.cocorahs.org/.

26. National Weather Service Forecast Office, Buffalo, NY, Buffalo SnowSpotter Network, http://www.erh.noaa.gov/buf/nws_buffalo_snowspotter_network.htm.

27. National Weather Service, Cooperative Observer Program, Snow Measurement Guidelines, http://www.nws.noaa.gov/om/coop/reference/Snow_Measurement_Guidelines _05-1997.pdf.

28. Christopher C. Burt and Mark Stroud, *Extreme Weather: A Guide and Record Book*, 2nd ed. (New York: W. W. Norton, 2007), 80. Stroud, a skilled freelance cartographer, received qualified co-authorship for producing the book's impressive maps and graphics.

29. The two events occurring outside US borders are a 70.9-inch snowfall within 15 hours at Dartmoor, Great Britain, on February 16, 1929, and a 67.8-inch snowfall within 19 hours at Bessans, France, on April 5–6, 1959; ibid.

30. The newspapers had the same owner, and the phrase quoted was the identical in both accounts; see Robert Gavin, "Series of Accidents Closes I-81 for Two Hours," *Syracuse Post-Standard*, December 23, 1993; and Scott Scanlon, "Snow Buries Parts of County," *Syracuse Herald-Journal*, Oswego edition, December 23, 1993.

31. Scanlon, "Snow Buries Parts of County."

32. NOAA, *Storm Data* 35, no. 12 (December 1993): 24.

33. US Department of Commerce, Environmental Science Services Administration, *Storm Data* 8, no. 2 (December 1966): 116.

34. *Climatological Data: New York* 78, no. 2 (December 1966): 196.

35. Robert B. Sykes, Jr., "Oswego's Tardy, Tough Winter of 1971," *Weatherwise* 6 (1972): 276–83; individual measurements on 280. Also see Robert B. Sykes, Jr., "1971/1972 Winter at Oswego, NY: A Short Commentary," *Proceedings of the 29th Eastern Snow Conference, Oswego, NY, February 3–5, 1972*, 32–46.

36. Quotation from NOAA, *Storm Data* 14, no. 1 (January 1972): 8. Also see *Climatological Data: New York* 84, no. 1 (January 1972): 4 and 15.

37. The four 30-minute measurements (4.3, 4.8, 4.2, and 4.2 inches) sum to exactly 17.5 inches.

38. NOAA, *Storm Data* 35, no. 1 (January 1993): 45–47; and *Climatological Data: New York* 105, no. 1 (January 1993): 21–25.

39. See the note for 1993 under January 18 in AccuWeather.com's Today in Weather History forum, http://forums.accuweather.com/. Several other websites perpetuate this record, including the DataStreme Daily Summary, sponsored by the University Consortium for Atmospheric Research and the American Meteorological Society; in particular, see the list "Historical Weather Events" included in the commentary for Monday, January 18, 1999, http://www.comet.ucar.edu/dstreme/archive/course/spr_99/s99w-1m_sum.html.

40. Charlie Wilson, "Weather History: January 18: Record Warm, Cold, Snowstorm, Ice, Wind, Storms, Tornadoes & Floods," Examiner.com, http://www.examiner.com/weather-in-wilmington/weather-history-january-18-record-warm-cold-snowstorm-ice-wind-storms-tornadoes-floods.

41. *Climatological Data: New York* 106, no. 1 (January 1994): 21–24.

42. Stephen W. Dill, "Snow Buried North Counties: A Lake-Effect Storm Drops Three Feet on Parts of Oswego and Jefferson Counties," *Syracuse Post-Standard*, January 20, 1994; and Paul J. Kocin, Daniel H. Graf, and William E. Gartner, "Snow," *Weatherwise* 48 (March 1995): 24–29; quotation on 27.

43. National Weather Service, *Evaluation of the Montague Snowfall*, 11.

44. *Climatological Data: New York* 71, no. 1 (January 1959): 4 and 14. The Bennetts Bridge snowfall was ninth on Burt's list, just below a fifteen-hour snowfall reported for Dartmoor, in England. His list also included a 62-inch snowfall within twenty-two hours at Freyer's Ranch, Colorado, on April 14–15, 1927.

45. National Climatic Data Center, Station Snow Climatology, http://nidis1.ncdc.noaa.gov/USSCViewer/.

46. Mary Pedley, e-mail communication, June 13, 2011.

47. Ed White, "Buffalo, Step Aside—When It Comes to Snow, the City in New York Has Nothing on Petoskey," *Grand Rapids Press*, January 3, 2002; and David A. Robinson and Sonya Senkowsky, "The 2001–02 Snow Season," *Weatherwise* 56 (March/April 2003): 38–45.

48. B. L. Wiggin, "Great Snows of the Great Lakes," *Weatherwise* 3 (1950): 123–26; quotation on 123. For a more contemporary account, see J. H. Spencer, "Exceptionally Severe Snowstorm of October 18–19, 1930, near Buffalo, N.Y.," *Monthly Weather Review* 58 (1930): 422.

49. Anne Mosher, e-mail communication, June 23, 2011.

50. Catie O'Toole and John Doherty, "Fulton DPW Digs Out Twice," *Syracuse Post-Standard*, January 22, 2008, Oswego edition.

51. Robert A. Baker, "Motorcycle Shop Owner Needs Home," *Syracuse Post-Standard*, February 28, 2010. Although the triggering event was a synoptic storm, not lake effect, the building had been weakened by the cumulative effects of years of snow loading.

52. Mike McAndrew, "City's Hard Bargain on Demolition," *Syracuse Post-Standard*, August 23, 2010.

53. Strong winds also exert a vertical lifting force on roofs. Stephen Patterson and Madan Mehta, *Roofing Design and Practice* (Upper Saddle River, NJ: Prentice Hall, 2001), 155–72.

54. See H. C. S. Thom, "Distribution of Maximum Annual Water Equivalent of Snow on the Ground," *Monthly Weather Review* 94 (1966): 265–71. Thom presented highly generalized national maps of the mean and standard deviation of the logarithms of snow water equivalent, based on only ten years of data for 140 stations. Although these local means and standard deviations could have been used to make a probability-based snow load map, Thom wisely recognized the limitations of his data and chose not to do so.

55. Bruce Ellingwood et al., *Development of a Probability-Based Load Criterion for American National Standard A58*, NBS Special Publication 577 (Washington, DC: National Bureau of Standards, 1980). The report influenced a range of applications, including the structural design of nuclear power plants. For a discussion of its development and impact, see Bruce Ellingwood, "Probability-Based Load Criteria in Structural Design," in *A Century of Excellence in Measurements, Standards, and Technology: A Chronicle of Selected NBS/NIST Publications 1901–2000*, NIST Special Publication 958, ed. David R. Lide, 283–88 (Washington, DC: National Institute of Standards and Technology, 2001), esp. 286–87.

56. For an overview of the implementation, see Bruce Ellingwood and Robert Redfield, "Ground Snow Loads for Structural Design," *Journal of Structural Engineering* 109 (1983): 950–64.

57. My principal informant on CRREL's development of the first and second snow load maps was Wayne Tobiasson, particularly in a March 17, 2010, telephone conversation and an April 14, 2010, e-mail. A civil engineer who specialized in snow loads—his 1974 master's thesis for the Thayer School of Engineering is titled "Stresses Developed on Structures Buried in Snow"—Tobiasson retired from CRREL in 1996 but continued to consult on issues of snow load.

58. Although rain gauges at these second-order stations recorded inches of precipitation, because of melting and runoff these measurements don't translate reliably into the weight of snow on the ground. To make their map, CRREL scientists had to find a statistical relationship between extreme snow depth and snow load for the first-order stations, and then use the formula to convert 2 percent snow depths at the second-order stations to 2 percent snow loads.

59. To make its new map CRREL developed a massive database, which it updated in the early 1990s. Measurements originally compiled for winters 1951–52 through 1979–80 were extended through 1991–92. Because useful data were not available for every station for every year, the 1982 map was based on roughly twenty years of record for each first-order station and a generally shorter period for most second-order stations. The update also increased the number of the first-order stations to 204, with an average record of thirty-three years, and raised the number of second-order stations to more than 9,200. See Wayne Tobiasson and Alan Greatorex, "Database and Methodology for Conducting Site Specific Snow Load Case Studies for the United States," in *Snow Engineering: Recent Advances*, ed. Masanori Izumi, Tsutomu Nakamura, and Ronald L Sack, 249–56 (Rotterdam: A. A. Balkema, 1997). Also see American Society of Civil Engineers, *Minimum Design Loads for Buildings and Other Structures* (Reston, VA: American Society of Civil Engineers, Structural Engineering Institute, 2010), esp. 29, 34–35, and 425–27.

60. According to Thomas Schmidlin and his colleagues, the 1982 and 1990 maps were essentially identical, with "the only changes . . . in the northern Great Plains." Thomas W. Schmidlin, Dennis J. Edgell, and Molly A. Delaney, "Design Ground Snow Loads for Ohio," *Journal of Applied Meteorology* 31 (1992): 622–27; quotation on 622. Wayne Tobiasson (e-mail April 14, 2010) specifically noted that the changes pertained only to Minnesota, North and South Dakota, and Montana.

61. Ronald L. Sack, "Designing Structures for Snow Loads," *Journal of Structural Engineering* 115 (1989): 303–15.

62. American Society of Civil Engineers, *Minimum Design Loads for Buildings and Other Structures* (New York: American Society of Civil Engineers, 1990), quotations on 26. The title page notes that ACSE 7-88, which was approved in December 1988 and published in July 1990, is a revision of ANSI A58.1-1982. Quotations are from the map key, on 26.

63. Michael O'Rourke, *Snow Loads: Guide to the Snow Load Provisions of ASCE 7-05* (Reston, VA: American Society of Civil Engineers, 2007), western half of fig. 7-1, on 146.

64. Although the map outlined general trends, architects and engineers need to read the fine print. Careful inspection of my excerpt (fig. 5.9) will reveal underlined zone numbers for several zones adjacent to a CS area. On the original CRREL map the labels for these zones include an elevation in parentheses above the zone number. According to the map key, "numbers in parentheses represent the upper elevation limits in feet for the ground snow load values presented [here]." These parenthetical notations are shown on my downsized excerpt (fig. 5.10, *upper right*), on which, for example, the "(1000)" atop the "40" for the zone that abuts the southern shore of Lake Ontario indicates that the 40 psf minimum ground snow load does not apply to places more than 1,000 feet above sea level. If CRREL had published its map at a much larger, more detailed scale, these places would appear as small, local patches of CS. As the map key warns, "Site-specific case studies are required to establish ground snow loads at elevations not covered." Topographic maps for this area reveal that places above 1,000 feet are generally isolated, few in number, and sparsely populated. Farther south, where elevations above 1,000 feet become more numerous or their patches large enough to put on the map, the mapped category changes to CS. Simply put, extensive highlands with steep slopes are too unpredictable for a generalized, moderately detailed snow load zone.

65. The Granite State was an ideal test bed because it is relatively small, and thus manageable, and 140 of its towns were in the CS zone on CRREL's second map. What's more, the laboratory is located there, in Hanover, just north of Dartmouth College. The procedure relies on local and regional data and addresses elevation differences with a statewide adjustment factor of 2.1 psf per 100 feet, deemed valid for elevations up to 2,500 feet—Mount Washington (elevation 6,288 ft.) and lesser peaks demand individual treatment and extreme caution. See Wayne Tobiasson et al., *Ground Snow Loads for New Hampshire*, ERDC/CRREL publication TR-02-6 (Hanover, NH: US Army Engineer Research and Development Center, 2002), quotations on 29 and 39.

66. New York State Fire Prevention and Building Code Council, *New York State Uniform Fire Prevention and Building Code* (New York: New York [State] Division of Housing and Community Renewal, 1984), 214–15.

67. James L. Harding, PE, Codes Division, New York (State) Department of State, e-mail, September 3, 2010.

68. New York (State) Division of Code Enforcement and Administration, *Building Code of New York State* (Falls Church, VA: International Code Council; Albany: New York [State] Department of State, 2002), 295.

69. The accompanying text attributed the zone boundaries to the *Atlas of Extreme Snow-Water Equivalent for the Northeastern United States* produced in 1994 by the Northeast Regional Climate Center at Cornell University, and noted that the Cornell map, based on "a 50-year return period, [was] found to correlate well with the ASCE 7 snow map." New York (State) Division of Code Enforcement and Administration, *Building Code of New York State* (Washington, DC: International Code Council; Albany:

New York [State] Department of State, 2007), 26. The *Atlas* map in question is "50-year Return Period: Seasonal Maximum Snow Water Equivalent," map 7 in Daniel S. Wilks and Megan McKay, *Atlas of Extreme Snow-Water Equivalent for the Northeastern United States*, Research Publication RR 94-3 (Ithaca, NY: Northeast Regional Climate Center, Cornell Univ., 1994), 18. Although the isolines on the *Atlas* map represent the water-equivalent of snow in inches while the boundary lines on the building-code map outline zones of minimum snow load, their delineations are highly similar. In general, an inch of water equivalent on the Atlas map equates to a snow load increment of roughly 6 psf, which is slightly more than 5.2 pounds, the weight of ½2 cubic foot of cold fresh water.

70. US Department of Defense, *Unified Facilities Criteria (UFC): Structural Load Data*, report UFC 3-310-01, 5 December 2007. A factor of 20.8854 was used to convert values in kilopascals (from page C-8 in table C-1) to pounds per square foot. This report superseded an earlier version, also numbered UFC 3-310-01 but dated June 30, 2000.

71. US Department of Defense, *Unified Facilities Criteria*, table 1, pages 1-3 through 1-4.

72. Thomas L. Smith, "Metal Roof Systems: Design Considerations for Snow and Ice," *Professional Roofing*, November 1991, 74; and Thomas L. Smith, "Part Two: Snow and Ice on Metal Roof Systems," *Professional Roofing*, December 1991, 62.

73. For insights on the advantages and resurgence of metal roofs, see Dan Chiras, "Eco-Friendly Roofing Options," *Mother Earth News* no. 240 (June/July 2010): 97, 99–101; Larry D. Hodge and Sally S. Victor, "Comeback of the Metal Roof," *Popular Science* 223, no. 4 (October 1983): 124–26, 200; Nancy Keates, "Homeowners on a Hot Metal Roof—Durable, Stylish Workhorse Makes a Comeback," *Wall Street Journal*, February 22, 2008; D. A. Taylor, "Sliding Snow on Sloping Roofs," *Canadian Building Digest* no. 228 (Division of Building Research, National Research Council of Canada, Ottawa, November 1983); and Wayne Tobiasson and James Buska, "Standing Seam Metal Roofing Systems in Cold Regions," *Proceedings, 10th Conference on Roofing Technology* (Rosemont, IL: National Association of Roofing Contractors, 1993), 34–44.

6. Change

1. Robert DeC. Ward, "The Snowfall of the United States," *Scientific Monthly* 9 (1919): 397–415; quotations on 413–15. For insights to Ward's impact as a scholar, see Robert V. Rohli and Gregory D. Bierly, "The Lost Legacy of Robert DeCourcy Ward in American Geographical Climatology," *Progress in Physical Geography* 35 (2011): 547–64.

2. Climate change is hardly a recent issue insofar as warming trends were hotly debated in colonial America, and theories about the meteorological impacts of increased anthropogenic CO_2 in the atmosphere can be traced at least as far back as the late

nineteenth century, most notably in the work of Swedish physicist Svante Arrhenius. See James Rodger Fleming, *Historical Perspectives on Climate Change* (New York: Oxford Univ. Press, 1998).

3. Ludlum founded the magazine *Weatherwise* in 1948, and his books include *Early American Winters*, published by the American Meteorological Society in two volumes, for 1604–1820 (1966) and 1821–1870 (1968). In an 1987 essay, for instance, Ludlum debunked eighteenth-century notions of a changing climate with milder winters but steered clear of contemporary notions of climate change; see David M. Ludlum, "The Climythology of America," *Weatherwise* 40 (1987): 255–59. For an obituary, see Robert McG. Thomas, Jr., "David Ludlum, Weather Expert, Dies at 86," *New York Times*, May 29, 1997.

4. Scanned images of observers' sheets, showing all measurements reported for each month, can be downloaded from the NDCD Free Data Web site, at http://www7.ncdc .noaa.gov/IPS/coop/coop.html.

5. *Break* is defined in item 6 in Data Inventory Indicators, section 1.4.2 of National Climate Data Center, "US Snow Climatology Background," http://www.ncdc.noaa.gov/ussc/USSCAppController?action=scproject.

6. Metadata summarizing the quality of co-op data are available online from the National Climate Data Center at http://www.ncdc.noaa.gov/ussc/USSCAppController?action=scproject.

7. National Climatic Data Center, "Lowville, New York—Top Ten Snowiest Years," http://www.ncdc.noaa.gov/ussc/USSCAppController?action=snowfall_topten&state=30 &station=LOWVILLE&coopid=304912&param=10.

8. This technique, known as least-squares linear regression, calculates a line that minimizes the sum of the squared vertical deviations between the points and the line.

9. The probability (p) is 0.0123. Because physical scientists generally consider values less than 0.05 "significant," the upward slope is deemed statistically meaningful. The related squared regression coefficient (r^2) is 0.07, indicating that the least-squares line accounts for 7 percent of the variation in snowfall. Although the "fit" between points and line is weak, the line's upward direction is unlikely to have arisen by chance.

10. The curved line was fit to the data by JMP-IN, a statistical software application distributed by SAS Institute, Inc. Specifically, the curve is a cubic spline fit to the data using a lambda of 0.05.

11. $p = 0.0003$ and $r^2 = 0.15$.

12. For 1924–44, $p = 0.08063$ and $r^2 = 0.16$, and for 1958–83, $p = 0.2922$ and $r^2 = 0.05$. Because the probabilities exceed 0.05, neither regression line is considered statistically significant.

13. For 1934–79, $p = 0.0002$ and $r^2 = 0.28$, and for 1979–2010, $p = 0.0127$ and $r^2 = 0.20$. Because the probabilities are less than 0.05, both regression lines are considered statistically significant.

14. Adam W. Burnett et al., "Increasing Great Lake-Effect Snowfall During the Twentieth Century: A Regional Response to Global Warming?" *Journal of Climate* 16 (2003): 3535–42.

15. Andrew W. Ellis and Jennifer J. Johnson, "Hydroclimatic Analysis of Snowfall Trends Associated with the North American Great Lakes," *Journal of Hydrometeorology* 5 (2004): 471–86.

16. For discussion of the quality of snowfall data, which is a potential source of spurious trends, see Doesken and Judson, *The Snow Booklet*, 57–60; and David A. Robinson, "Evaluation of the Collection, Archiving and Publication of Daily Snow Data in the United States," *Physical Geography* 10 (1989): 120–30.

17. Daria Scott and Dale Kaiser, "Changes in Characteristics of United States Snowfall over the Last Half of the Twentieth Century," paper presented on February 12, 2003, at the 14th Symposium on Global Change and Climate Variations, held in Long Beach, California, and sponsored by the American Meteorological Society. An extended abstract is online at http://ams.confex.com/ams/pdfpapers/53981.pdf.

18. Daria Scott and Dale Kaiser, "Variability and Trends in United States Snowfall over the Last Half Century," paper presented on January 13, 2004, at the 15th Symposium on Global Change and Climate Variations, held at Seattle, WA. An extended abstract, which includes additional maps, is online at http://ams.confex.com/ams/pdfpapers/71795.pdf.

19. Jerome Namias, "Snowfall over Eastern United States: Factors Leading to Its Monthly and Seasonal Variations," *Weatherwise* 13 (1960): 238–47; quotation on 240.

20. Eichenlaub, "Lake Effect Snowfall to the Lee of the Great Lakes," 403–12.

21. D. C. Norton and S. J. Bolsenga, "Spatiotemporal Trends in Lake Effect and Continental Snowfall in the Laurentian Great Lakes, 1951–1980," *Journal of Climate* 6 (1993): 1943–56.

22. Daniel J. Leathers and Andrew W. Ellis, "Synoptic Mechanisms Associated with Snowfall Increases to the Lee of Lakes Erie and Ontario," *International Journal of Climatology* 16 (1966): 1117–35. Also see Andrew W. Ellis and Daniel J. Leathers, "A Synoptic Climatological Approach to the Analysis of Lake-Effect Snowfall: Potential Forecasting Applications," *Weather and Forecasting* 11 (1996): 216–29.

23. Kenneth E. Kunkel et al., "A New Look at Lake-Effect Snowfall Trends in the Laurentian Great Lakes Using a Temporally Homogeneous Data Set," *Journal of Great Lakes Research* 35 (2009): 23–29; quotation on 29.

24. For fuller discussion of the use of expert evaluations to screen for data quality, see Kenneth E. Kunkel et al., "Trends in Twentieth-Century US Snowfall Using a Quality-Controlled Dataset," *Journal of Atmospheric and Oceanic Technology* 26 (2009): 33–44.

25. For discussion of the broader influences of El Niño and La Niña on regional snowfall patterns (in contrast to lake-effect snowfall alone), see Kenneth E. Kunkel et

al., "Trends in Twentieth-Century US Extreme Snowfall Seasons," *Journal of Climate* 22 (2009): 6204–16.

26. Shawn R. Smith and James J. O'Brien, "Regional Snowfall Distributions Associated with ENSO: Implications for Seasonal Forecasting," *Bulletin of the American Meteorological Society* 82 (2001): 1179–91. For an example of El Niño's uncertain relationship to snowfall in western and central New York, see National Weather Service Forecast Office, Buffalo, NY, "El Niño's Influence on WNY's Winter Weather," http://www.erh.noaa.gov /buf/research/climate/elnino_main.htm.

27. For a concise introduction, see NASA, Goddard Space Flight Center, Global Master Change Directory, Monthly Oceanic Nino Index (ONI), http://gcmd.nasa.gov /records/GCMD_NOAA_NWS_CPC_ONI.html.

28. See American Meteorological Society, *Glossary of Meteorology*, online at http:// amsglossary.allenpress.com/glossary.

29. I've greatly simplified Grimaldi's findings and discussion; for additional details, see Richard Grimaldi, "Climate Teleconnections Related to El Niño Winters in a Lake-Effect Region of West-Central New York," *Atmospheric Science Letters* 9 (2008): 18–25.

30. Only the January 22 to February 22 period had a statistically significant correlation, though. Richard Grimaldi, e-mail communication, March 3, 2011. Grimaldi's recent research focuses on the role of the North Atlantic Oscillation, whereby a warmer-than-normal phase of the North Atlantic Ocean, characterized by warmer surface waters, fewer icebergs, and more open sea, operates through a general shift in atmospheric pressure to produce colder temperatures and wind patterns favorable to lake-effect snow.

31. Daria B. Kluver, "Characteristics and Trends in North American Snowfall from a Comprehensive Gridded Data Set" (master's thesis, Univ. of Delaware, 2007), esp. 109, 113–14, and 133–34.

32. M. Notaro, W-C. Wang, and W. Gong, "Model and Observational Analysis of the Northeast US Regional Climate and Its Relationship to the PNA and NAO Patterns during Early Winter," *Monthly Weather Review* 134 (2006): 3479–3505.

33. Adam W. Burnett, telephone conversation, February 25, 2011.

34. Burnett is one of the co-authors of Henry T. Mullins et al., "Holocene Climate and Environmental Change in Central New York (USA)," *Journal of Paleolimnology* 45 (2011): 243–56.

35. Peter J. Sousounis and George M. Albercook, "Potential Futures," in *Preparing for a Changing Climate: The Potential Consequences of Climate Variability and Change—Great Lakes*, Great Lakes Regional Assessment Group, 19–24 (Ann Arbor: Univ. of Michigan, 2000).

36. Kenneth E. Kunkel, Nancy E. Westcott, and David A. R. Kristovich, "Climate Change and Lake-Effect Snow," in *Preparing for a Changing Climate: The Potential Consequences of Climate Variability and Change—Great Lakes*, Great Lakes Regional Assessment Group, 25–27 (Ann Arbor: Univ. of Michigan, 2000).

37. A. James Wagner, "The Record-Breaking Winter of 1976–77," *Weatherwise* 30 (1977): 65–69.

38. R. A. Wrightson, "The Wild Winter of 1976–77 in New York State," *Weatherwise* 30 (1977): 70–75.

39. Kenneth F. Dewey, "Lake-effect Snowstorms and the Record Breaking 1976–77 Snowfall to the Lee of Lakes Erie and Ontario," *Weatherwise* 30 (1977): 228–31.

40. For studies of impacts in the Great Lakes region, see George W. Kling et al., *Confronting Climate Change in the Great Lakes Region: Impacts on Our Communities and Ecosystems* (Cambridge, MA: Union of Concerned Scientists; Washington, DC: Ecological Society of America, 2003); and Peter J. Sousounis, "The Future of Lake-Effect Snow: A SAD Story," published in January/February 2000 in *Acclimations*, newsletter of US National Assessment of the Potential Consequences of Climate Variability and Change, and online at the Global Change Research Program website, http://www.usgcrp.gov/ usgcrp/nacc/greatlakes.htm. Also see Peter C. Frumhoff, *Confronting Climate Change in the U.S. Northeast: Science, Impacts, and Solutions* (Cambridge, MA: Union of Concerned Scientists, 2007), esp. chapter 6; and NYSERDA ClimAID Team, *Responding to Climate Change in New York State* (Albany: New York State Energy Research and Development Authority, 2010), esp. 4.

41. Union of Concerned Scientists, Great Lakes Communities and Ecosystems at Risk Project, Interactive migrating climates Web page, http://www.ucsusa.org/greatlakes /glimpactmigrating.html. For a recent application of migrating climates to states bordering Lake Michigan, see Katharine Hayhoe et al., "Regional Climate Change Projections for Chicago and the US Great Lakes," *Journal of Great Lakes Research* 36 (2010): 7–21. An earlier graphic based on a "higher-emissions scenario" placed Upstate New York in central Georgia in 2070–2090. See Peter C. Frumhoff et al., *Confronting Climate Change in the U.S. Northeast: Science, Impacts, and Solutions* (Cambridge, MA: Union of Concerned Scientists, 2007), 7.

42. Lessened dependence on Middle Eastern oil would surely reduce US military involvement in the region, and a reduced use of oil and coal would yield cleaner air. For a relatively balanced assessment of the global warming donnybrook, I recommend Greg Craven, *What's the Worst That Could Happen?: A Rational Response to the Climate Change Debate* (New York: Penguin Books, 2009). Craven, a high school science teacher adept at boiling down complicated issues, presents a clear, highly informative overview. For insight to the clashes between climate scientists and their doubting critics, see Howard Friel, *The Lomborg Deception: Setting the Record Straight about Global Warming* (New Haven, CN: Yale Univ. Press, 2010). I also admire the assertive recognition of climate change in Paul Gross, *Extreme Michigan Weather: The Wild World of the Great Lakes State* (Ann Arbor: Univ. of Michigan Press, 2010), esp. 48–63. Gross, a meteorologist at a Detroit television station, offers a thoughtful, impressively concise overview of key issues.

7. Place

1. The area became known as the Burned-over District because of frequent "conflagrations of religious excitement." Michael Barkun, *Crucible of the Millennium: The Burned-over District of New York in the 1840s* (Syracuse: Syracuse Univ. Press, 1986), 2–12; quotation on 3.

2. Richard L. Bushman, *Joseph Smith and the Beginnings of Mormonism* (Urbana: Univ. of Illinois Press, 1984), 4, 31, 56, and 61–62.

3. Basinger, Jeanine, *The* It's a Wonderful Life *Book* (New York: Alfred A. Knopf, 1986), 23–27.

4. For a litany of similarities between Bedford Falls and Seneca Falls, see "The Real Bedford Falls. . . . Too Many Coincidences to Ignore," http://therealbedfordfalls.com /therealbedfordfalls.php.

5. "'It's A Wonderful Life' in Seneca Falls, Friday, December 10th–Sunday, December 12th, 2010," http://therealbedfordfalls.com/events.php. Schedule of events changes yearly.

6. For information about snowmobiling, see http://www.gosnowmobiling.org, the website of the International Snowmobile Manufacturers Association, and *The New York State Snowmobiler's Guide*, prepared by the Snowmobile Unit of the New York State Department of Parks, Recreation, and Historic Preservation (Albany, 2010), online at http://nysparks.state.ny.us/recreation/snowmobiles/documents/SnowmobilersGuide.pdf.

7. The website http://www.SnowMobileLewisCounty.com touts the county's 500 miles of trail and provides links to thirteen area clubs, six dealers, five rental firms, and five repair shops.

8. Lewis County, New York, Comprehensive Plan, October 4, 2008, chapter 2, 41. The plan is available online at http://www.lewiscountyny.org/content/Generic/View/20.

9. National Climatic Data Center, New York Snowfall and Snow Depth Extremes Table, http://www.ncdc.noaa.gov/ussc/USSCAppController?action=extremes&state=30.

10. Welcome to Louie's!, http://www.hodkinsonsgrill.com/index.htm.

11. US Geological Survey, Geographic Names Information System, online at http:// geonames.usgs.gov/.

12. John and Urial Hooker are the only county residents with the surname Hooker mentioned in Franklin B. Hough, *A History of Lewis County in the State of New York from the Beginning of Its Settlement to the Present Time* (Albany, NY: Munsell and Rowland, 1860). Hooker is in the town of Montague, but Hough's brief account of the town's history (193–94) mentions neither Hooker the place nor Hooker the person.

13. The NCDC's Multi-Network Metadata System, which links to Google Maps, is online at https://mi3.ncdc.noaa.gov/mi3qry/login.cfm.

14. Welcome to Montague Inn, http://www.montague-inn.com/.

15. Kelly Vadney, "Work of a Wind Tower Explained," *Watertown Daily Times*, January 7, 2007; and Steve Virkler, "Maple Ridge Pays Communities $8.94M," *Watertown Daily Times*, January 4, 2011.

16. Federal Writers' Project of the Works Progress Administration, *New York: A Guide to the Empire State* (New York: Oxford Univ. Press, 1940), 641.

17. In early 2011, another large blaze opened another gap; see Steve Virkler, "Fire Destroys Lowville Block," *Watertown Daily Times*, February 6, 2011.

18. FindTheBest.com lists numerous other not-for-profit community organizations in Lowville, at http://non-profit-organizations.findthebest.com/directory/d/New-York /LOWVILLE.

19. Federal Writers' Project, *New York: A Guide to the Empire State*, 641; Martha Ellen, "Dairy Plants on the Decline," *Watertown Daily Times*, January 31, 2011; and Lewis County Chamber of Commerce, Cheese in Lewis County, http://www.adirondacks tughill.com/history_cheese.php.

20. "Lowville AMF May Gain Jobs," *Watertown Daily Times*, January 15, 2008; and Richard Barnett, "Bowling Pin Plant Closes, Leaving only One in U.S.," *Augusta (Georgia) Chronicle*, January 20, 2008.

21. Paul Mackun and Steven Wilson, "Population Distribution and Change: 2000 to 2010," 2010 Census Briefs, US Census Bureau, March 2011.

22. Richard L. Florida, *The Rise of the Creative Class, and How It's Transforming Work, Leisure, Community, and Everyday Life* (New York: Basic Books, 2002). Florida, who parleyed his theory into a profitable sideline as a circuit lecturer and consultant, is not without critics. For examples, see Stefan Krätke, "'Creative Cities' and the Rise of the Dealer Class: A Critique of Richard Florida's Approach to Urban Theory," *International Journal of Urban and Regional Research* 34 (2010): 835–53; and Alex Macgillis, "The Ruse of the Creative Class," *American Prospect* 21 (January/February 2010): 12–16.

23. *Businessweek* provided concise summaries, with selected Census results, at http:// images.businessweek.com/slideshows/20110525/america-s-best-affordable-places-2011 /slides/23 (for Marquette County, Michigan) and . . . slides/33 (for Onondaga County, New York). Educational attainment is for persons 25 and older. For comparable nationwide tabulations, see US Census Bureau, State & County Quick Facts, http://quickfacts .census.gov/qfd/states/00000.html.

24. See the 40 Below website, http://www.40belowsummit.com/.

25. Quotations from Ngoc Huynh and Emily Kulkus, "'Cuse Comebacks: Be a Syracuse Ambassador and Respond to Slams on the City," *Syracuse Post-Standard*, June 1, 2007, as posted on the 40 Below website, at http://www.40belowsummit.com/news /showNews.php?n=106.

26. Julia Terruso, "Cold Hands, Warm Arts: Upstate Snowdown Fundraiser Features Outdoor Activities," *Syracuse Post-Standard*, February 6, 2011.

27. Marnie Eisenstadt, "Just as We Thawed: Winterfest Always Seems to Guarantee a Warm Spell, Right? Well, This Heat Wave Was Short-Lived," *Syracuse Post-Standard*, February 19, 2011; and Abram Brown, "Chili Cook-Off Heats Winterfest—Judges Favor Up in Smoke, While DJ's Inn Wins Public's Vote," *Syracuse Post-Standard*, February 27, 2011.

28. Anne Mosher, e-mail communication, June 23, 2011.

29. Eichenlaub, *Weather and Climate of the Great Lakes Region*, 93–102.

30. Anne Mosher noted a similar trade-off in California, where treacherous fog "gives us great wine."

Illustration Credits

Frontis. Adapted from C. L. Mitchell, "Snow Flurries along the Eastern Shore of Lake Michigan," *Monthly Weather Review* 49 (1921): 502.

1.1. Redrawn from Val L. Eichenlaub, "Lake Effect Snowfall to the Lee of the Great Lakes: Its Role in Michigan," *Bulletin of the American Meteorological Society* 51 (1970): 404.

1.2. Adapted from map on the Chestnut storm web page, National Weather Service, Buffalo office.

1.3. Image from Chestnut storm web page, National Weather Service, Buffalo office.

1.4. Adapted from color map on the Chestnut storm web page, National Weather Service, Buffalo office.

1.5. Adapted from image on the Jet Propulsion Laboratory's photojournal web page.

1.6. Data compiled by author.

2.1. Excerpted from map in "Third Report of the Committee on Meteorology," *Journal of the Franklin Institute* 19 (1837): 19.

2.2. Redrawn from map in Franklin B. Hough, *Results of a Series of Meteorological Observations Made in Obedience to Instructions from the Regents of the University, at Sundry Academies in the State of*

New York, from 1826 to 1850 Inclusive (Albany: Weed, Parsons and Company, 1855), before title page.

2.3.   Upper map redrawn from "Hyetal or Rain Chart: Mean Distribution of Precipitation for the Winter," in US Surgeon-General's Office, *Meteorological Register, for Twelve Years, from 1843 to 1854, Inclusive, from Observations Made by the Officers of the Medical Department of the Army, at the Military Posts of the United States* (Washington, DC: A. O. P. Nicholson, 1855), folded map between 734 and 735. Lower map redrawn from "Hyetal or Rain Chart: Mean Distribution of Rain for the Winter on the North American Continent," in Lorin Blodget, *Climatology of the United States, and of the Temperate Latitudes of the North American Continent, Embracing a Full Comparison of These with the Climatology of the Temperate Latitudes of Europe and Asia, and Especially in Regard to Agriculture, Sanitary Investigations, and Engineering, with Isothermal and Rain Charts for Each Season, the Extreme Months, and the Year* (Philadelphia: J. B. Lippincott and Co., 1857), folded map between 242 and 243.

2.4.   Adapted from two-color (black and blue) map in Charles A. Schott, *Tables and Results of the Precipitation, in Rain and Snow, in the United States: and at Some Stations in Adjacent Parts of North America, and in Central and South America*, Smithsonian Contributions to Knowledge 222 (Washington, DC: Smithsonian Institution, 1872), back of volume.

2.5.   Redrawn at a smaller size from the "Map of the State of New York Showing the Average Precipitation for the Winter," printed with black and blue inks, in New York Meteorological Bureau and Weather Service, *Fifth Annual Report*, 1893 (Albany: James B. Lyon, 1894), back of volume.

2.6.   From the eight-map Snowfall sheet, printed in orange and dark blue inks, in Mark W. Harrington, *Rainfall and Snow of the United States, Compiled to the End of 1891, with Annual, Seasonal, Monthly, and Other Charts*, Bulletin C—Atlas (Washington, DC: Weather Bureau, 1894), sheet xviii.

2.7.   Upper-left and upper-right excerpts redrawn at a smaller size from maps on the Snowfall sheet in Mark W. Harrington, *Rainfall and Snow of the United States, Compiled to the End of 1891, with Annual, Seasonal, Monthly, and Other Charts*, Bulletin C—Atlas (Washington, DC: Weather Bureau, 1894), sheet xviii. Lower-left excerpt redrawn from Frank Waldo, *Elementary Meteorology for High Schools and Colleges* (New York: American Book Company, 1896), 344. Lower-right excerpt redrawn at a smaller size from Chart xi, Average Annual Snowfall, accompanying A. J. Henry, "Normal Annual Sunshine and Snowfall," *Monthly Weather Review* 26 (1898): 108.

2.8.   Excerpted from map in Charles F. Brooks, "The Snowfall of the United States," *Quarterly Journal of the Royal Meteorological Society* 39 (1913): foldout between 82 and 83. Map key was repositioned from lower left.

2.9.   Excerpted from map in the atlas accompanying Charles Franklin Brooks, "The Snowfall of the Eastern United States" (PhD diss., Harvard University, 1914), chart XV.

2.10.   Excerpt redrawn at a smaller scale from "Mean Annual Snowfall (Inches)," in Brooks and Connor, *Climatic Maps of North America* (Cambridge, MA: Blue Hill Meteorological Observatory of Harvard University and Harvard University Press, 1936), map 21.

2.11.   Reproduced photographically at a slightly smaller scale from color map in J. B. Kincer, Advance sheets set 5, Part II, Climate; Section A, Precipitation and Humidity, *Atlas of American Agriculture* (Washington, DC: Government Printing Office, 1922), 43.

2.12.   Upper map excerpted from figure 76 in J. B. Kincer, Advance sheets 5, part 2, Climate; section A, Precipitation and Humidity, *Atlas of American Agriculture* (Washington, DC: Government Printing Office, 1922), 42. Lower map excerpted from map in J. B. Kincer, "Climate and Weather Data for the United States," in US Department of Agriculture, *Climate and Man*, 1941 Yearbook of Agriculture (Washington, DC: Government Printing Office, 1941), 727.

2.13. Excerpts redrawn from the two maps on the full-page Snowfall sheet in US Geological Survey, *National Atlas of the United States* (Washington, DC, 1970), 100.

2.14. Upper map adapted from map in US Environmental Data Service, *Climatic Atlas of the United States* (Washington, DC: Government Printing Office, 1968), 53. Lower map adapted from the color map "Lower 48 States SNOW—Mean Total Snowfall (Annual)" on the National Climatic Data Center's Climate Maps of the United States web page.

2.15. Adapted from color map in John H. Thompson, ed., *Geography of New York State* (Syracuse, NY: Syracuse University Press, 1966), fig. 22 in pocket.

2.16. Excerpted from map in Richard P. Cember and Daniel S. Wilks, *Climatological Atlas of Snowfall and Snow Depth for the Northeastern United States and Southeastern Canada*, Research Series publication no. RR 93-1 (Ithaca, NY: Northeast Regional Climate Center, 1993), map 18.

2.17. Excerpted from figure 1-b in Robert W. Scott and Floyd A. Huff, "Impacts of the Great Lakes on Regional Climate Conditions," *Journal of Great Lakes Research* 22 (1996): 849.

3.1. Adapted from C. L. Mitchell, "Snow Flurries along the Eastern Shore of Lake Michigan," *Monthly Weather Review* 49 (1921): 502.

3.2. R. M. Dole, "Snow Squalls of the Lake Region," *Monthly Weather Review* 56 (1928): 512.

3.3. Lewis F. Richardson, *Weather Prediction by Numerical Process* (Cambridge: Cambridge University Press, 1922), 184.

3.4. Forecast-skill trend lines are from a graph on a NOAA website celebrating the agency's 200th birthday. Related technological improvements are mostly from Frederick G. Shuman, "History of Numerical Weather Prediction at the National Meteorological Center," *Weather and Forecasting* 4 (1989): 286–96.

3.5. Compiled by author from the AMS Journals Online website.

3.6. Adapted from maps in James E. Jiusto, Douglas A. Payne, and Michael L. Kaplan, *Great Lakes Snowstorms, Part 2: Synoptic and Climatological Aspects* (Albany, NY: Atmospheric Sciences Research Center, State University of New York, 1970), 41, 44, and 45.

3.7. Adapted from maps in Ronald L. Lavoie, "A Mesoscale Numerical Model of Lake-Effect Storms," *Journal of the Atmospheric Sciences* 29 (1972): 1032 and 1037.

3.8. Compiled from state-level maps available through the NWS Directives System web page.

3.9–10. Redrawn from map in NEXRAD Panel, National Weather Service Modernization Committee, National Research Council, *Toward a New National Weather Service: Assessment of NEXRAD Coverage and Its Associated Weather Services* (Washington, DC: National Academies Press, 1995), 27.

3.11. From US General Accounting Office, *National Weather Service: Modernization Activities Affecting Northwestern Pennsylvania*, Report no. GAO/AIMD-97-156, September 1997, 22.

3.12. Adapted from color maps on the Graphical Forecasts–Eastern Great Lakes web page of the National Weather Service, Buffalo office.

3.13. Adapted from color graphics on the Experimental Probabilistic Snowfall Forecast for the Upper Peninsula web page of the National Weather Service, Marquette office.

3.14. Adapted from color map on the Experimental Probabilistic Snowfall Forecast for the Upper Peninsula web page of the National Weather Service, Marquette office.

4.1.   Excerpt redrawn from map in Stanley A. Changnon, David
        Changnon, and Thomas R. Karl, "Temporal and Spatial
        Characteristics of Snowstorms in the Contiguous United States,"
        Journal of Applied Meteorology and Climatology 45 (2006): 1144.

4.2.   From color map on NOAA's NESIS web page.

4.3.   Adapted from color map on the Buffalo Weather Forecast Office
        website.

4.4.   Adapted from color map, 7.8 inches wide, in Rochester Bureau of
        Municipal Research, *Report on the Problem of Snow Removal in the
        City of Rochester, N.Y.* (Rochester, 1917), facing 32.

4.5.   Photograph by Mark Monmonier.

4.6–9. Compiled from summaries posted online by local television stations.

5.1.   Photograph by Mark Monmonier.

5.2.   Photograph by Grant Goodge, a member of the Montague snowfall
        committee, included in National Weather Service, Office of
        Meteorology, *Evaluation of the Reported January 11–12, 1997,
        Montague, New York, 77-inch, 24-hour Lake-effect Snowfall* (Silver
        Spring, MD: National Oceanic and Atmospheric Administration,
        1997), 12.

5.3.   From National Weather Service, Office of Meteorology, *Evaluation
        of the Reported January 11–12, 1997, Montague, New York, 77-inch,
        24-hour Lake-effect Snowfall* (Silver Spring, MD: National Oceanic
        and Atmospheric Administration, 1997), 5.

5.4.   Compiled by author.

5.5.   *Syracuse Post-Standard.* Reproducible black-and-white version
        constructed by author using similar symbols and type.

5.6.   Compiled by author from the NCDC's New York Snowfall and Snow Depth Extremes Table.

5.7.   Compiled by author from the NCDC's Michigan Snowfall and Snow Depth Extremes Table.

5.8.   Excerpt adapted from maps in American Society of Civil Engineers, *Minimum Design Loads for Buildings and Other Structures* (New York: American Society of Civil Engineers, 1990), 25–26.

5.9.   Excerpt adapted from map in Michael O'Rourke, *Snow Loads: Guide to the Snow Load Provisions of ASCE 7-05* (Reston, VA: American Society of Civil Engineers, 2007), 146–47.

5.10.   Upper-left map redrawn from map in New York State Fire Prevention and Building Code Council, *New York State Uniform Fire Prevention and Building Code* (New York: New York [State] Division of Housing and Community Renewal, 1984), 215. Upper-right excerpt adapted from map in Michael O'Rourke, *Snow Loads: Guide to the Snow Load Provisions of ASCE 7-05*, 146–47. Lower-left map redrawn from map in New York (State) Division of Code Enforcement and Administration, *Building Code of New York State* (Falls Church, VA: International Code Council; Albany, NY: New York State Department of State, 2002), 296. Lower-right map redrawn from map in New York (State) Division of Code Enforcement and Administration, *Building Code of New York State* (Washington, DC: International Code Council; Albany, NY: New York [State] Department of State, 2007), 27.

5.11.   Photograph by author.

6.1–4. Compiled from data provided by the National Climatic Data Center and the Northeast Regional Climate Center.

6.5.   Based on map and table in Adam W. Burnett et al., "Increasing Great Lake-Effect Snowfall during the Twentieth Century: A Regional Response to Global Warming?" *Journal of Climate* 16 (2003): 3537–38.

6.6.   Redrawn and generalized from maps in Andrew W. Ellis and Jennifer J. Johnson, "Hydroclimatic Analysis of Snowfall Trends Associated with the North American Great Lakes," *Journal of Hydrometeorology* 5 (2004): 482. The map projection was adjusted so that north–south scale more closely approximates east–west scale; overall map scale was decreased, and the isoline interval for the right-hand map was doubled to reduce the number of isolines.

6.7.   Adapted from figures 1 and 2 in Daria Scott and Dale Kaiser, "Changes in Characteristics of United States Snowfall over the Last Half of the Twentieth Century," paper presented on February 12, 2003, at the 14th Symposium on Global Change and Climate Variations, held at Seattle, Washington.

6.8.   Adapted from color maps on the NOAA Climate Prediction Center website.

6.9.   Redrawn from the Migrating Climates map for Michigan on the Great Lakes Communities and Ecosystems at Risk—The Impacts: Migrating Climates web page of the Union of Concerned Scientists.

6.10.   Redrawn from the Migrating Climates map for New York on the Great Lakes Communities and Ecosystems at Risk—The Impacts: Migrating Climates web page of the Union of Concerned Scientists.

7.1.   Compiled by author from CLIMOD Daily Lister data provided by the Northeast Regional Climate Center for station 308383, Syracuse Hancock International Airport.

7.2.   Excerpt adapted from the lower-48-states snow-days color map on the National Climatic Data Center's Climate Maps of the United States web page.

7.3.   Excerpt adapted from the sunrise-to-sunset sky-cover/visibility (cloudy-days) map for January on the National Climatic Data Center's Climate Maps of the United States web page.

7.4.   Layout and typography by author.

7.5.   Excerpt adapted from the color map "High Snowfall Areas of the NYS Funded Snowmobile Trail System, 2010–2011" on the New York State Department of Parks, Recreation, and Historic Preservation website.

7.6–14. Photographs by Mark Monmonier.

7.15.   Excerpt adapted from percentage-change map in Paul Mackun and Steven Wilson, "Population Distribution and Change: 2000 to 2010," 2010 Census Briefs, US Census Bureau (March 2011), 7.

Index